**Springer Theses**

Recognizing Outstanding Ph.D. Research

## Aims and Scope

The series "Springer Theses" brings together a selection of the very best Ph.D. theses from around the world and across the physical sciences. Nominated and endorsed by two recognized specialists, each published volume has been selected for its scientific excellence and the high impact of its contents for the pertinent field of research. For greater accessibility to non-specialists, the published versions include an extended introduction, as well as a foreword by the student's supervisor explaining the special relevance of the work for the field. As a whole, the series will provide a valuable resource both for newcomers to the research fields described, and for other scientists seeking detailed background information on special questions. Finally, it provides an accredited documentation of the valuable contributions made by today's younger generation of scientists.

## Theses are accepted into the series by invited nomination only and must fulfill all of the following criteria

- They must be written in good English.
- The topic should fall within the confines of Chemistry, Physics, Earth Sciences, Engineering and related interdisciplinary fields such as Materials, Nanoscience, Chemical Engineering, Complex Systems and Biophysics.
- The work reported in the thesis must represent a significant scientific advance.
- If the thesis includes previously published material, permission to reproduce this must be gained from the respective copyright holder.
- They must have been examined and passed during the 12 months prior to nomination.
- Each thesis should include a foreword by the supervisor outlining the significance of its content.
- The theses should have a clearly defined structure including an introduction accessible to scientists not expert in that particular field.

More information about this series at http://www.springer.com/series/8790

Benedikt Frieß

# Spin and Charge Ordering in the Quantum Hall Regime

Doctoral Thesis accepted by
the Technical University of Munich, Germany

Springer

*Author*
Dr. Benedikt Frieß
Max Planck Institute for Solid State Research
Stuttgart
Germany

*Supervisor*
Prof. Klaus von Klitzing
Max Planck Institute for Solid State Research
Stuttgart
Germany

ISSN 2190-5053 ISSN 2190-5061 (electronic)
Springer Theses
ISBN 978-3-319-81539-8 ISBN 978-3-319-33536-0 (eBook)
DOI 10.1007/978-3-319-33536-0


Softcover reprint of the hardcover 1st edition 2016


Printed on acid-free paper

This Springer imprint is published by Springer Nature
The registered company is Springer International Publishing AG Switzerland

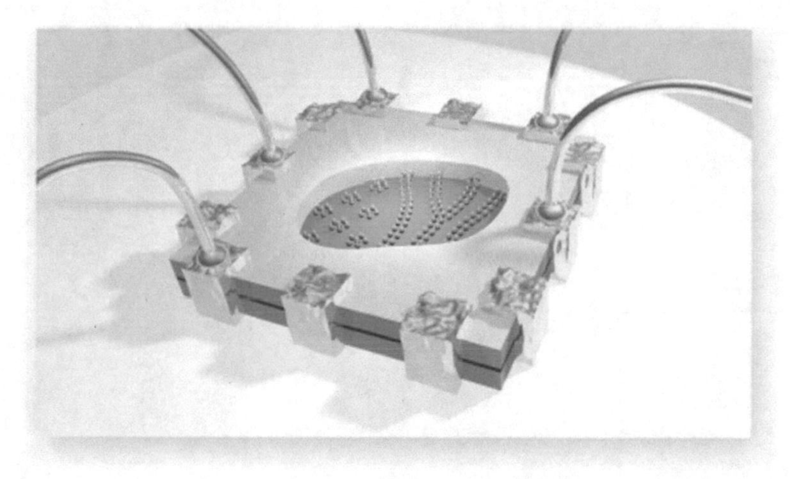

Artistic rendering of the spin and charge
ordering in a semiconductor heterostructure

# Supervisor's Foreword

Quantum Hall physics started more than 35 years ago and is still an expanding field, thanks to new high-quality materials such as graphene and zinc oxide, new theoretical concepts including topological insulators and topological quantum computation, and last but not least, due to important applications in metrology that will lead to a new international system of units based on constants of nature.

The most important ingredient in this research field is merely a two-dimensional electron system (2DES) in a strong perpendicular magnetic field leading (in an ideal system) to a discrete energy spectrum. The 2DES is not a mathematical simplification of a three-dimensional system; it truly opens a new dimension in science, especially if magnetic flux quanta are added. New quasiparticles called composite bosons, composite fermions, or anions are realized with new electronic properties. In some cases, quasiparticles with non-abelian statistics are expected that may well revolutionize quantum information technology. The host material in this research field of basic science plays a secondary role and the influence of the crystal on the 2DES is normally included in parameters such as the effective mass and the g-factor of the electrons. Unfortunately, even high-quality crystals are never perfect, and their residual defects and impurities very often mask the intrinsic properties of the electronic system. A two-dimensional electron layer on the clean surface of liquid helium is very nearly an ideal system for basic science in the two-dimensional world, but the electron concentrations in these experiments are limited to very small values so that the diversity of quantum Hall physics cannot be exploited in these systems.

At present, the best-controlled 2DES for basic research is realized at the interface between GaAs and AlGaAs, and every time the quality of this system (characterized by the mobility of electrons) has been increased, new phenomena were observed.

The integer quantum Hall effect can be discussed within a simple independent-electron picture, where the discrete energy levels for electrons in a strong magnetic field (Landau levels) are fully occupied and an incompressible electron system is formed. Disorder does not destroy this effect. Quite the contrary, it stabilizes this quantum phenomenon. However, many other fascinating electron phases based on electron–electron interactions such as the fractional quantum Hall effect, Wigner crystals, or exciton condensation become visible only if the disorder is negligibly small. Two other prominent phenomena based on correlated electrons, namely the formation of bubbles and stripes, are the focus of the research performed by Benedikt Frieß as described in this thesis.

Contrary to the intuitive picture of electrons in a plane with Coulomb repulsion, the expected homogeneous electron concentration can be drastically changed by redistributing the electrons if, for example, the increase in Coulomb energy is overcompensated by the gain in exchange energy. The electrons in a half-filled Landau level may redistribute in stripes representing a mixture of fully occupied Landau levels with alternating numbers of completely filled energy levels. The resulting electronic anisotropy can clearly be seen in angular-dependent resistivity measurements, but another technique used by Benedikt Frieß, the analysis of the velocity change of surface acoustic waves due to the screening properties of the electron system, yields unexpectedly new information about the compressibility of the spatially varying electron system. Such density-modulated phases are of general interest because they exist in nature in many cases, including some high-$T_C$ superconductors. Contrary to other systems, the control of the electron concentration in GaAs/AlGaAs heterostructures offers a unique opportunity to study these phases in detail. Very sensitive probes are available for characterizing the density-modulated phases because the charge-modulated system is generally connected to a spin-density modulation, which can be detected sensitively by local probes via the Knight shift.

This thesis provides not only a comprehensive introduction to modern aspects of quantum Hall physics but also an impressive overview of experimental details and techniques for characterizing quantum Hall systems with a focus on new methods based on electron–nuclear spin interactions.

Stuttgart
January 2016

Prof. Dr. Dr. h.c. mult.
Klaus von Klitzing

# Acknowledgments

The present thesis would have lacked much of its profundity if it weren't for the great support I have received from many people. In particular, I am highly indebted to...

- Professor Klaus von Klitzing for giving me the privilege to become one of his students and for providing exceptional support in many regards. I highly value the many insightful discussions and constructive suggestions. His enthusiasm and supportive mind has been greatly inspiring to me.
- Dr. habil. Jurgen Smet for guiding me through all the ups and downs of the recent years. His extensive support combined with all necessary liberties provided best possible research conditions and pushed me to excel. I could not have wished for a better environment.
- my external supervisor Prof. Bernd Rosenow (Universität Leipzig) for the numerous helpful discussions. The present work strongly benefited from his wealth of experience and profound knowledge.
- Professor Rudolf Gross (TU München), who kindly agreed to become the second referee of my doctoral exam.
- Dr. Vladimir Umansky (Weizmann Institute of Science) for providing some of the world's best 2DES.
- the cleanroom team for bringing the 2DES into shape. Most of all, I would like to thank Marion Hagel for her persistent help in sample processing.
- our skillful technicians, Steffen Wahl, Gunther Euchner, Steffen Pischke, Ingo Hagel and Manfred Schmid, for their invaluable help when it comes to cooling the samples to ultra-low temperatures.
- Dr. Johannes Nübler for introducing me to the broad sweep of low-temperature physics. I learned a lot from him, from the intricacies of lab work to the details of quantum Hall physics.
- all my friends and colleagues in the Smet group. The joy you brought me in the past years beggars description. I highly appreciate the many scientific, philosophical and mundane discussions. Most prominently I would like to mention Matthias Kühne, Ding Zhang, Federico Paolucci, Daniel Kärcher, Thomas Weitz, Youngwook Kim, Johannes Geurs and Patrick Herlinger.

- Joseph Falson (University of Toyko, now at MPI) for bringing the rich physics of ZnO/MgZnO heterostructures and himself to Stuttgart.
- Professor Daniel Loss (Universität Basel) for inviting me to Basel and sharing his knowledge about nuclear magnetic ordering.
- Professor Yongqing Li (Chinese Academy of Sciences) for fruitful collaboration on the field of topological insulators and quantum Hall spin physics. I am truly grateful for the invitation to Beijing.
- Dr. Yang Peng and Prof. Felix von Oppen (both FU Berlin) for helping us to develop a theoretical understanding of the surface acoustic wave experiments
- Christian Reichl and Prof. Werner Wegscheider (ETH Zürich) for providing additional high-quality samples.
- Ruth Jenz for keeping the bureaucracy at bay and for her friendly nature.
- the financial and non-material support by the Studienstiftung des deutschen Volkes during my undergraduate studies.
- my friends and family for their invaluable support throughout my life. Most of all, I feel deeply obliged to my parents. They sparked in me the excitement for nature and technology early on and taught me to think critically.
- Julia for enriching my daily life. Thank you for being so patient and understanding through thick and thin and for cheering me up when things went wrong in the lab.

# Contents

# Symbols and Abbreviations

## Symbols

| | |
|---|---|
| $\alpha$ | SAW phase |
| $\gamma$ | Gyromagnetic ratio |
| $\epsilon$ | Electric permittivity |
| $\lambda$ | Wavelength |
| $\mu$ | Electron mobility |
| $\mu_0$ | Vacuum permeability |
| $\mu_B$ | Bohr magneton |
| $\mu_M$ | Nuclear magnetic moment |
| $\mu_N$ | Nuclear magneton |
| $\nu$ | Filling factor |
| $\tilde{\nu}$ | Filling factor in the topmost occupied Landau level |
| $\varrho$ | Resistivity |
| $\rho$ | Particle density |
| $\sigma$ | Conductivity |
| $\tau$ | Relaxation and scattering time |
| $\phi$ | Electrostatic potential |
| $\Psi$ | Electron wavefunction |
| $\omega_c$ | Cyclotron frequency |
| $\Gamma$ | Landau-level broadening |
| $\Delta_\nu$ | Energy gap at filling factor $\nu$ |
| $\Phi = h/e$ | Magnetic flux quantum |
| $c$ | Stiffness tensor |
| $e$ | Elementary charge |
| $f$ | Frequency |
| $g_e^*$ | Effective electron g-factor |
| $g_N$ | Nuclear spin g-factor |
| $h = 2\pi\hbar$ | Planck constant |
| $i$ | Subband index |

| | |
|---|---|
| $j$ | Current density |
| $k$ | Wavevector |
| $l_B$ | Magnetic length |
| $m$ | Electron rest mass |
| $m^*$ | Electron effective mass |
| $m_j$ | Nuclear magnetic quantum number |
| $m_p$ | Proton rest mass |
| $n$ | Landau level index |
| $n_e$ | Electron density |
| $n_L$ | Landau level degeneracy |
| $p$ | Even integer number |
| $\overline{p}$ | Pressure |
| $\mathcal{P}$ | Piezoelectric constant |
| $q$ | Odd integer number |
| $t$ | Time |
| $u$ | Displacement amplitude |
| $w$ | Quantum well width |
| $\boldsymbol{A}$ | Vector potential |
| $A_{HF}$ | Hyperfine interaction constant |
| $\boldsymbol{B}$ | Magnetic flux density |
| $B^*$ | Effective magnetic field |
| $D$ | Electrical displacement field |
| $\boldsymbol{E}$ | Electric field |
| $E_c$ | Cyclotron energy |
| $E_F$ | Fermi energy |
| $E_Z$ | Zeeman energy |
| $I$ | Electrical current |
| $J$ | Nuclear spin angular momentum |
| $\mathcal{J}$ | Nuclear spin quantum number |
| $\mathrm{K}$ | Electromechanical coupling coefficient |
| $K_s$ | Knight shift |
| $L$ | Length |
| $N_A$ | Avogadro constant |
| $\mathcal{P}$ | Electron spin polarization |
| $\mathcal{P}^*$ | Electron spin polarization normalized by $n_L$ |
| $\mathcal{P}_N$ | Nuclear spin polarization |
| $R$ | Resistance |
| $\overline{R}$ | Universal gas constant |
| $R_c$ | Cyclotron radius |
| $R_H$ | Hall resistance |
| $R_{xx}$ | Longitudinal resistance |
| $S$ | Electron spin angular momentum |
| $\mathcal{S}$ | Surface area |
| $T$ | Temperature |

$T_1$ Nuclear spin relaxation time
$\tau$ Stress tensor
$V$ Voltage
$\overline{V}$ Volume
$W$ Width

## Abbreviations

1D One-dimensional
2DES Two-dimensional electron system(s)
AC Alternating current
AlGaAs Aluminium gallium arsenide
BET Brunauer–Emmett–Teller
CDW Charge density wave
CF Composite fermion
DC Direct current
DOS Density of states
FQHE Fractional quantum Hall effect
FQHS Fractional quantum Hall state(s)
GaAs Gallium arsenide
IQHE Integer quantum Hall effect
IQHS Integer quantum Hall state(s)
MBE Molecular beam epitaxy
NMR Nuclear magnetic resonance
QW Quantum well
RF Radio frequency
RIQHS Reentrant integer quantum Hall state(s)
SAW Surface acoustic waves
SPSL Short-period superlattice

# Chapter 1
# Introduction

The motion of electrons in a three-dimensional solid state system is generally characterized by numerous degrees of freedom spanning a large phase space. Despite the vast number of microscopic states, the overall macroscopic behavior of such systems is often rather simple and can in most cases be conveniently described by only a few quantities. It appears almost paradoxical that reducing the electronic degrees of freedom, in contrast, yields some of the most diverse and fascinating phenomena in physics. Constraining the number of allowed states often entails otherwise more subtle interactions to play a dominant role. The quantum Hall effect, studied in this thesis, serves as a splendid example of such behavior. It is stage to a multitude of intriguing phenomena formed by the interplay of competing many-body interactions. To facilitate the emergence of the quantum Hall effect and the diverse physics within, constraints must be put on the electrons in different regards.

The first important prerequisite for the existence of the quantum Hall effect is the confinement of electrons in one direction. A versatile platform to create such two-dimensional electron systems (2DES) is provided by semiconductor heterostructures. Here, the electrons are kept in place by an impenetrable barrier formed at the interface of semiconductors with different band gaps. This system offers a good controllability of the confinement strength as well as an excellent quality of the 2DES. In recent years, also material systems which naturally form a two-dimensional electron system have become of major interest (e.g. graphene).

In addition to the confinement of electrons in the direction perpendicular to the 2DES, the number of allowed states needs to be reduced further. This is done by restricting the movement also in the plane of the 2DES, not spatially as before but by putting constraints on the kinetic energy. For this purpose, a magnetic field is applied perpendicular to the 2DES. Quantum mechanics dictates that the kinetic energy of a 2DES condenses into a set of discrete values if exposed to a strong magnetic field. As a result, the electronic energy spectrum is composed of highly degenerate, discrete levels separated by the cyclotron energy—the so-called Landau levels. The degeneracy of each Landau level depends on the magnetic field strength. Thus, the filling fraction of the Landau levels, characterized by the filling factor $\nu$, constantly changes when increasing the magnetic field. This important property of a 2DES lays



B. Frieß, *Spin and Charge Ordering in the Quantum Hall Regime*,
Springer Theses, DOI 10.1007/978-3-319-33536-0_1

the foundation of the famous (integer) quantum Hall effect. It manifests itself in standard electron transport experiments as plateaus in the Hall resistance at integer values of the filling factor combined with a vanishingly small longitudinal resistance, provided that the temperature is low enough to suppress thermal excitations to higher Landau levels [1].

The integer quantum Hall effect (IQHE), whose mere existence can be understood on the basis of non-interacting particles, forms the framework for a remarkable collection of interaction-driven phenomena. As all electrons within a Landau level have the same kinetic energy and excitations to higher Landau levels are frozen out, the physics at partial filling is dominated by the interplay between Coulomb, Zeeman and exchange interactions. This situation is most evident in state-of-the-art GaAs/AlGaAs heterostructures—up to date the superior platform to study many-body interactions in the quantum Hall regime. At high magnetic fields, when the lowest Landau level is partially occupied, fractional quantum Hall states (FQHS) emerge [2]. These quantum Hall states at fractional filling factors share the appearance of the IQHE but are of a different origin. They are manifestations of a collective state which arises from many-particle Coulomb interactions. At certain commensurate ratios of the charge and magnetic flux density, the electrons can lower their Coulomb energy by binding to magnetic flux quanta. This correlated state gives rise to a fascinating and counterintuitive breed of quasiparticles, so-called anyons. They bear fractional charges and exhibit an exchange statistic different from fermions and bosons. As a clear sign of correlated physics, the existence of these quasiparticles cannot be understood by extrapolating the behavior of single electrons.

When looking at higher Landau levels, beyond filling factor $\nu = 4$, the situation is drastically different. No fractional quantum Hall states form here. Instead, density-modulated phases are prevalent at partial fillings. They arise by the spontaneous ordering of electrons in spatial patterns with either one-dimensional stripe order (stripe phase) or a two-dimensional crystalline order (bubble phase) [3]. Such phases constitute an evident signature of systems with a strong competition between attractive and repulsive interactions acting on different length scales. Similar physics is at play at the formation of stripe and bubble phases in magnetic and organic films [4]. In the quantum Hall regime, these phases emerge from the competition between the repulsive direct Coulomb interaction and the attractive exchange interaction [3, 5]. Their dominance over the fractional quantum Hall effect stems from subtle differences in the relative strength of these interactions, primarily caused by changes in the shape and extent of the wavefunction in higher Landau levels. The competition between density-modulated phases and fractional quantum Hall states manifests itself most clearly in the second Landau level. Here, both phases are of almost equal strength, and minute changes of the electron density or the magnetic field may induce transitions between them.

The diverse many-body phenomena featured in the quantum Hall regime are enriched even further by the spin degree of freedom [6]. It leads to an additional splitting of the Landau levels (Zeeman effect) with sometimes far-reaching consequences. The fractional quantum Hall states, for instance, are known to exist with different degrees of spin polarization depending on the ratio of Coulomb and

Zeeman energy. By tuning the relative strength of both energies, transitions between the respective spin phases can be induced. These spin transitions are accompanied by the occurrence of ferromagnetic domains [7–9]. Even outside the fractional quantum Hall regime, complex spin structures do exist such as skyrmionic spin formations around filling factor $\nu = 1$ [10, 11] and a spatially ordered spin density in the regime of the density-modulated phases.

By the wealth of examples given above, the quantum Hall effect appears in metaphorical terms as a veritable goldmine in the field of many-body physics. To many of the intriguing phenomena we shall return in the course of this thesis. Special emphasis is thereby put on the spin physics as well as the density-modulated phases in the quantum Hall regime. The topics covered in detail are outlined below.

- Chapter 2 treats the basic properties of a two-dimensional electron system at low temperatures with and without an external magnetic field. At first, details on the formation of a 2DES in a GaAs/AlGaAs quantum well structure are given. This is followed by a section about the basic equations of motion in two dimensions. The main part of the second chapter covers general aspects of the integer and fractional quantum Hall effect. Special attention is paid to the density-modulated phases in the quantum Hall regime and the exceptional physics of the second Landau level.
- Chapter 3 is about electron-nuclear spin interactions in the quantum Hall regime. The coupling of the electronic and nuclear spin system is an often unconsidered aspect when dealing with the quantum Hall effect. In most cases this is justified as it only leads to minor adjustments of the energy scales involved. Here, we explicitly utilize this coupling to systematically study the low-energetic spin excitations in the quantum Hall regime by measuring the nuclear spin relaxation rate. Surprisingly, we find that not only the nuclear spin relaxation is modified at certain filling factors but also the equilibrium value of the nuclear spin polarization. This observation is studied more carefully in the second part of this chapter.
- Chapter 4 addresses the long-lasting search for the spin polarization of the wondrous $\nu = 5/2$ state. For many years, the origin of the FQHS at $\nu = 5/2$ has attracted a lot of attention since it cannot be explained by the standard picture which accounts for most of the FQHS [12]. Over time, alternative theories have been developed, some of which predict a novel kind of quasiparticle to be present at the $\nu = 5/2$ state, so-called non-abelian anyons [13, 14]. The prospect of finding such quasiparticles has triggered substantial efforts to unravel the nature of this exceptional FQHS. An important question in this regard is the spin polarization of the $\nu = 5/2$ state. We have tackled this issue by means of nuclear magnetic resonance spectroscopy. This technique relies on the characteristic shift of the nuclear resonance frequency in the presence of a spin-polarized electron system (Knight shift). In order to affect the quality of the fragile $\nu = 5/2$ state as little as possible, we have employed a sensitive detection scheme based on changes in the sample resistance under resonant nuclear spin excitation.
- Chapter 5 deals with the microscopic structure of the stripe phases in the quantum Hall regime. Thus far, the density-modulated phases in the quantum Hall regime were studied mostly in transport experiments providing only macroscopic access.

Their microscopic nature remains a largely unexplored facet. Here, we address this topic by employing nuclear spins as local sensors to probe the density distribution of the stripe phase. For this purpose, as in the previous chapter, the Knight shift is utilized. A theoretical model is implemented to interpret the observed resonance behavior. The modulation strength and stripe period can be extracted with the help of this model.

- Chapter 6 remains in the field of density-modulated phases but focuses on a different aspect. Here, the bubble and stripe phases in the quantum Hall regime are studied by means of surface acoustic waves. These sound waves dynamically sense the conductivity and the screening behavior of the 2DES with the benefit of a well-defined probing direction. This property proves to be advantageous especially in the context of density-modulated phases since here the distribution of current flow is influenced decisively by the spatial density inhomogeneities.
- Chapter 7 provides a summary of the thesis.
- Appendix A specifies the details of the samples studied in the course of this thesis. It includes information on the layer sequence and lithographical patterning.
- Appendix B describes the design of the low-temperature sample holder manufactured specifically for the demanding needs of the experiments performed here. The density-modulated phases and most of all the fragile $\nu = 5/2$ state require ultra-low temperatures to exist. Special measures were taken to optimize the cooling power and to minimize the heat load even in the presence of electrical leads and microwave excitation.

## References

1. K. von Klitzing, G. Dorda, M. Pepper, New method for high-accuracy determination of the fine-structure constant based on quantized Hall resistance. Phys. Rev. Lett. **45**, 494 (1980)
2. D.C. Tsui, H.L. Stormer, A.C. Gossard, Two-dimensional magnetotransport in the extreme quantum limit. Phys. Rev. Lett. **48**, 1559 (1982)
3. M.M. Fogler, Stripe and bubble phases in quantum Hall systems, in *High Magnetic Fields: Applications in Condensed Matter Physics and Spectroscopy* (Springer, Berlin, 2002), pp. 98–138
4. M. Seul, D. Andelman, Domain shapes and patterns: The phenomenology of modulated phases. Science **267**, 476 (1995)
5. M.O. Goerbig, P. Lederer, C. Morais Smith, Competition between quantum-liquid and electron-solid phases in intermediate Landau levels. Phys. Rev. B **69**, 115327 (2004)
6. Y.Q. Li, J.H. Smet, Nuclear-electron spin interactions in the quantum Hall regime, in *Spin Physics in Semiconductors* (Springer, Berlin, 2008), pp. 347–388
7. J. Eom, H. Cho, W. Kang, K.L. Campman, A.C. Gossard, M. Bichler, W. Wegscheider, Quantum Hall ferromagnetism in a two-dimensional electron system. Science **289**, 2320 (2000)
8. J.H. Smet, R.A. Deutschmann, W. Wegscheider, G. Abstreiter, K. von Klitzing, Ising ferromagnetism and domain morphology in the fractional quantum Hall regime. Phys. Rev. Lett. **86**, 2412 (2001)
9. O. Stern, N. Freytag, A. Fay, W. Dietsche, J.H. Smet, K. von Klitzing, D. Schuh, W. Wegscheider, NMR study of the electron spin polarization in the fractional quantum Hall effect of a single quantum well: Spectroscopic evidence for domain formation. Phys. Rev. B **70**, 075318 (2004)

10. S.L. Sondhi, A. Karlhede, S.A. Kivelson, E.H. Rezayi, Skyrmions and the crossover from the integer to fractional quantum Hall effect at small Zeeman energies. Phys. Rev. B **47**, 16419 (1993)
11. S.M. Girvin, Spin and isospin: Exotic order in quantum Hall ferromagnets. Phys. Today **3**, 39 (2000)
12. R.L. Willett, The quantum Hall effect at 5/2 filling factor. Reports Prog. Phys. **76**, 076501 (2013)
13. G. Moore, N. Read, Nonabelions in the fractional quantum Hall effect. Nucl. Phys. B **360**, 362 (1991)
14. M. Greiter, X.G. Wen, F. Wilczek, Paired Hall states. Nucl. Phys. B **374**, 567 (1992)

# Chapter 2
# The Two-Dimensional Electron System

At the heart of this thesis are the properties of high-quality two-dimensional electron systems when being exposed to strong magnetic fields and low temperatures. In our case, the 2DES is hosted inside of a GaAs/AlGaAs heterostructure—a system well known for its excellent quality. Details on the sample structure are provided in the first section of this chapter. The remaining sections summarize the basic properties of a 2DES in a perpendicular magnetic field and lay the foundation for the physics discussed in subsequent chapters. Of particular importance is here the integer and fractional quantum Hall effect. Special emphasis is put on the density-modulated phases in the quantum Hall regime as well as the peculiar physics of the second Landau level. We restrict ourselves to the situation in GaAs/AlGaAs heterostructures. The following sections are based on introductory textbooks [1, 2] as well as Ph.D. theses [3–8].

## 2.1 Realizing a 2DES in a GaAs/AlGaAs Heterostructure

The confinement of electrons into two dimensions can be realized in a variety of different systems, such as semiconductor heterostructures, graphene and topological insulators. The samples investigated in the course of this thesis employ a GaAs/AlGaAs heterostructure to form a 2DES of exceptional quality. Up to date, it is the superior platform to study quantum Hall physics, showing a large number of integer and fractional quantum Hall states (Sects. 2.3 and 2.4) as well as different density-modulated phases (Sect. 2.5). The samples were fabricated by molecular beam epitaxy (MBE), a method which provides precise control over the vertical layer growth and excellent crystalline purity. Figure 2.1 shows the basic sample structure. The key features are described below. Further details about the samples used in the following chapters can be found in Appendix A. The wafers for fabricating these



B. Frieß, *Spin and Charge Ordering in the Quantum Hall Regime*,
Springer Theses, DOI 10.1007/978-3-319-33536-0_2

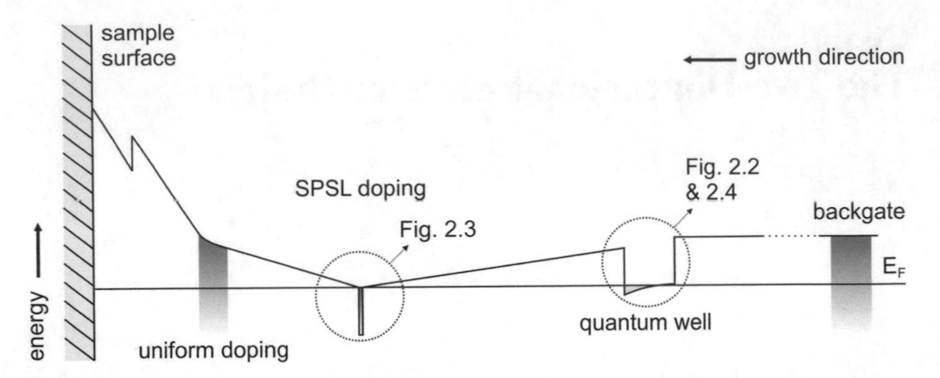


**Fig. 2.1** Schematic of the conduction band profile along the growth direction in a typical GaAs/AlGaAs quantum well structure (not drawn to scale). Modulation doping is used to populate the QW with electrons. The backgate is realized by incorporating a highly doped layer underneath the QW

samples of excellent quality were developed and kindly supplied by Dr. Vladimir Umansky (Weizmann Institute of Science, Israel).

**The Quantum Well**

The 2DES is located inside of a quantum well (QW) structure, which is formed by a thin layer of GaAs (typically 10–50 nm) sandwiched between AlGaAs barriers (Fig. 2.2). Because GaAs has a lower conduction band energy than AlGaAs, the movement of electrons is restricted to the GaAs layer provided that their energy is lower than the potential barrier of the QW. In order to fulfill this requirement, the QW width $w$, the density of electrons $n_e$ as well as their temperature must be chosen accordingly. The confinement in the vertical direction ($z$-axis) leads to discrete values of the $z$-component $k_z$ of the wavevector. In the case of a QW with infinitely high barriers, these values are $k_z = i\pi/w$ (i = 1, 2, 3, ...). Consequently, the energy associated with the movement of electrons in the growth direction is quantized to

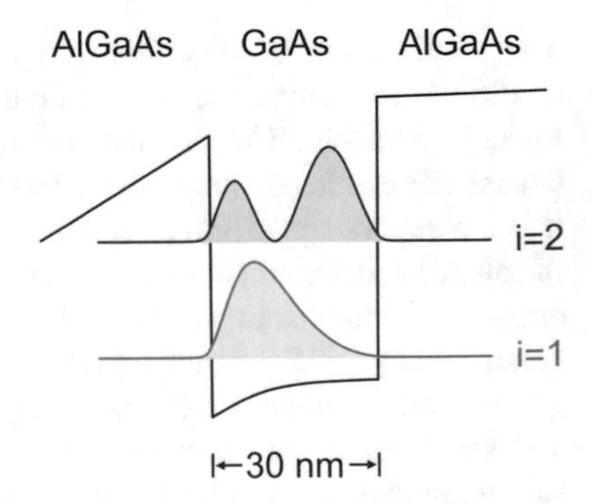


**Fig. 2.2** Probability density of the two lowest wavefunctions in a 30 nm-wide GaAs/AlGaAs quantum well

$$E_z = \frac{\hbar^2 k_z^2}{2m^*} = \frac{\hbar^2 \pi^2 i^2}{2m^* w^2}, \tag{2.1}$$

where $m^*$ is the effective mass of the conduction band electrons. Therefore, decreasing the QW width results in a larger spacing between the different energy subbands. In the more realistic case of a QW with finite barriers, the wavefunction may extend partially into the barriers, which increases the effective width of the QW. Figure 2.2 shows the probability density of the two wavefunctions which are lowest in energy for the case of a 30 nm-wide QW. The calculations were done by solving the Schrödinger and Poisson equations self-consistently using the software *nextnano++* [9]. The samples studied in this thesis are designed in such a way that only the lowest subband is occupied. In this case, all electrons have the same $k_z$ component, and the electron system can be considered two-dimensional despite the finite extent of the wavefunction along the z-direction (Fig. 2.2). For a 2DES with parabolic energy dispersion and non-interacting electrons, the number of states per unit area and energy can be calculated according to

$$\text{DOS} = \frac{m^*}{\pi \hbar^2}. \tag{2.2}$$

Thus, the density of states (DOS) of a two-dimensional electron system with parabolic dispersion depends merely on the effective electron mass and is in particular independent of the electron energy.

**The Doping Scheme**

Intrinsic GaAs (as well as AlGaAs) turns insulating at low temperatures. Therefore, dopants must be integrated into the crystal structure to populate the QW with electrons. To achieve a high-quality 2DES, it is essential how the doping is implemented. The dopants should affect the movement of the electrons in the QW as little as possible. The obvious approach, i.e. to dope directly the GaAs layer, is unfavorable because it degrades the mobility of the electrons due to scattering from ionized donors. Instead, we use the concept of modulation doping, which separates the doping layer from the QW by a distance of typically 50–100 nm [11]. Despite the spatial separation, electrons from the donor atoms will accumulate in the QW because of its (initially) lower-lying unoccupied energy levels (Fig. 2.1). Once the equilibrium state has been reached, the QW is filled with electrons up to the Fermi energy $E_F$.[1]

To realize a high-mobility 2DES, a large density of electrons is desirable because it mitigates the effect of disorder in the QW. The main sources of disorder in our heterostructures are residual impurities in the QW and the smooth disorder potential created by the remote ionized donors randomly distributed in the doping layer [10, 12]. The more electrons are located in the QW the better this disorder landscape is screened. The electron density in the QW can be adjusted by changing the distance of the doping layer from the QW. This distance is referred to as the spacer thickness. However, decreasing the spacer thickness increases the strength of the electrostatic

[1]More precisely, the electrochemical potential should be used instead of the Fermi energy at finite temperatures. We neglect this subtlety throughout the thesis in view of the low sample temperatures.

disorder in the QW due to the remote ionized donors and thus enhances the scattering rate of electrons in the QW. Alternatively, higher $n_e$ can be realized by increasing the number of doping atoms. This handle is of course limited to the point when the energy level of the donors coincides with the Fermi level. Once this situation is reached, increasing the doping density will not raise the electron density any further. Nevertheless, such a strong doping is often intentionally applied as it ensures a more homogeneous electron density across the wafer. In our samples, the doping strength is even increased beyond this point. The idea behind such an over-doping is that the additional electrons will remain in the doping layer and partly screen the disorder potential of the ionized donors. This effect has been proven crucial to improve the sample quality and in particular the strength of the fractional quantum Hall states [13]. On the other hand, over-doping can create a conducting layer at the site of the dopants, which gives rise to parallel conduction in transport measurements and may impede the stable operation of topgates as they are commonly used in interferometer structures.

In our samples, this issue is partly solved by using the dedicated doping scheme depicted in Fig. 2.3 and described in detail in references [10, 14, 15]. It is based on a short-period superlattice (SPSL) of thin GaAs layers separated by AlAs layers. The Si-dopants are only placed inside of the GaAs layers. Each of the neighboring AlAs layers forms a barrier at the $\Gamma$-point, leading to a confinement of the electrons like in a narrow quantum well. The same is true for the AlAs layers at the $X$-point. The confinement raises the energy of the electrons and improves the charge transfer into the 2DES. Furthermore, the superlattice is designed such that the ground state energy in the AlAs layers at the $X$-point is lower than the one of the GaAs layer. Thus, the excess electrons present due to over-doping have the character of $X$-electrons in AlAs, which implies a higher effective mass compared to the $\Gamma$-electrons in GaAs. The main idea behind this doping scheme is that the heavier electrons in the AlAs layers do not contribute as strongly to parallel conduction as electrons in a GaAs layer but are still mobile enough to partly screen the disorder potential generated by the ionized

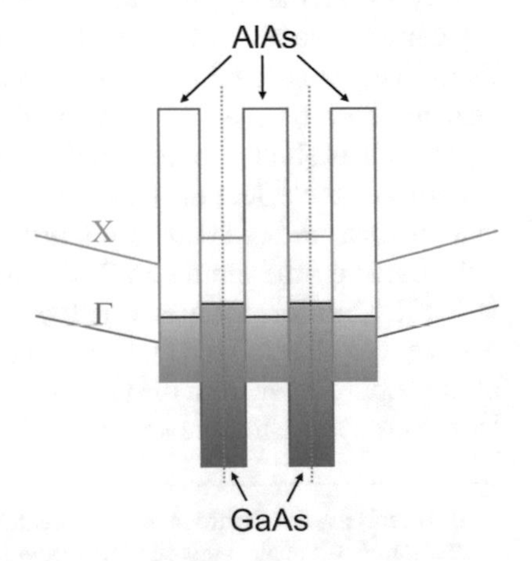


**Fig. 2.3** Conduction band energy at the $\Gamma$- and $X$-point for a short-period superlattice doping scheme (based on [10]). The doping is placed in the center of the GaAs layers (*dotted lines*)

donors. Doping directly the AlAs layer is intentionally avoided because it leads to the creation of DX centers. These negatively charged donor states cause lattice distortions of the crystal when trapping an additional electron and create states deep within the band gap [1, 16]. As a consequence, the charge transfer efficiency to the 2DES is reduced. Nevertheless, DX center doping is commonly used in high mobility 2DES structures because it gives rise to an effect known as persistent photoconductivity: Illuminating the sample with infrared light at low temperatures can release the trapped charges and increase the carrier concentration in the 2DES. After interrupting the illumination, the charge density remains high because an energy barrier has to be overcome for re-trapping the electrons in the DX centers. Above the SPSL doping structure a second, uniform doping layer is placed to compensate the influence of the surface states. Separating these two doping regions is not essential but simplifies the search for optimal doping parameters.

**The Backgate**

In many cases, the previously introduced doping scheme is used symmetrically on either side of the quantum well to double the charge density. In most of our samples, however, modulation doping is applied only at the top and left out at the bottom. Instead, a highly conductive layer is incorporated during growth at a distance of about 800 nm from the QW. This layer, if separately contacted, can be used to electrostatically change the electron density in the QW by applying different gate voltages. This feature is an important advantage. In fact, many of the experiments described in the following chapters would not have been possible without in-situ and fast control over the electron density. When tuning the backgate voltage, not only $n_e$ but also the shape of the electron wavefunction is affected. Figure 2.4 shows the shape of the wavefunction calculated for three different densities. In the present configuration, increasing the electron density obviously broadens the shape of the wavefunction. This behavior will become important in subsequent chapters.

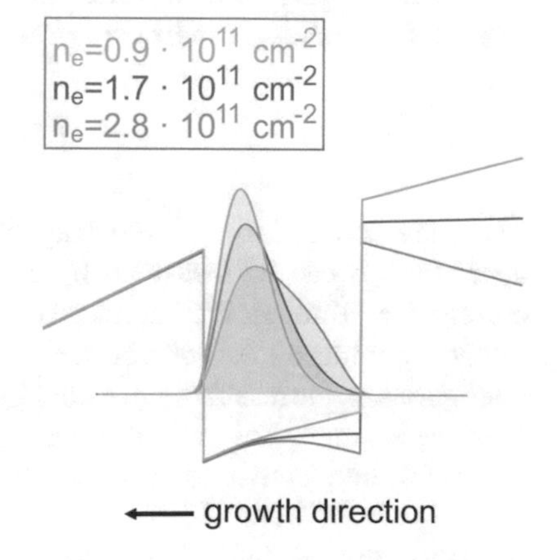


**Fig. 2.4** Probability density of the ground state wavefunction in a 30 nm-wide GaAs/AlGaAs quantum well. The density in the QW can be tuned electrostatically by a backgate. At the same time, changing the backgate voltage affects the shape of the wavefunction

## 2.2 Classical Electron Transport

The electron transport in two dimensions has many facets and depends on different parameters, such as sample quality, temperature and magnetic field. This section deals with the classical motion of electrons in two dimensions. We distinguish the situation with and without external magnetic field. The quantum-mechanical description of electron transport in two dimensions will be addressed in the next section. We limit the discussion to purely electronic transport.

**Transport Without Magnetic Field**

To approach the classical transport behavior, we employ the Drude model [17, 18]. In this model electrons are considered as point-like, massive particles. They get accelerated by an electric field $\boldsymbol{E}$. After a mean time $\tau$ electrons are scattered, for example by an impurity or a phonon. The scattering rate is treated as homogeneous and isotropic. Interactions between electrons are not taken into account. Furthermore, all electrons are assumed to contribute equally to the conductivity. Based on these assumptions, the equation of motion can be written as

$$m^* \frac{d\boldsymbol{v}}{dt} = -\frac{m^* \boldsymbol{v}}{\tau} - e\boldsymbol{E}, \tag{2.3}$$

where $\boldsymbol{v}$ denotes the velocity of the electron and $e$ the positive elementary charge. This leads in the stationary state $d\boldsymbol{v}/dt = 0$ to a constant drift velocity

$$\boldsymbol{v}_D = -\frac{e\tau}{m^*} \cdot \boldsymbol{E}. \tag{2.4}$$

The mobility $\mu = e\tau/m^*$ is therefore proportional to the scattering time $\tau$. The quality of a 2DES is often assessed by its mobility. In recent years, mobilities as high as $3 \times 10^7 \mathrm{cm}^2/\mathrm{Vs}$ have been realized [10]. This corresponds to a mean free path of more than 200 μm between consecutive scattering events. Using Eq. 2.4, the current density $\boldsymbol{j} = -n_e e \boldsymbol{v}_D$ can be expressed as

$$\boldsymbol{j} = \frac{n_e e^2 \tau}{m^*} \cdot \boldsymbol{E}. \tag{2.5}$$

Thus, the conductivity at zero magnetic field $\sigma_0$ can be written as $\sigma_0 = n_e e \mu = n_e e^2 \tau / m^*$. It can be measured by applying a constant voltage and detecting the current flow through the sample. However, for the experiments discussed in this thesis, we impose a constant current and measure the resulting voltage drop. In this configuration, the resistivity $\varrho$ of the sample is measured. In the absence of a magnetic field, $\varrho_0$ is simply the inverse of the conductivity $\varrho_0 = 1/n_e e \mu = m^*/n_e e^2 \tau$. Both quantities, conductivity and resistivity, can be conveniently used to determine the mobility of a 2DES provided that the electron density $n_e$ is known. $n_e$ can be extracted from the classical Hall effect with the help of an external magnetic field as described below.

### Transport Under the Influence of a Magnetic Field

When applying a magnetic field $B$ to the sample, the Lorentz force $\boldsymbol{F} = -e\boldsymbol{v} \times \boldsymbol{B}$ acts on the electrons and deflects them from a straight movement. As a consequence, the current flow $\boldsymbol{j}$ is no longer in line with the electric field direction $\boldsymbol{E}$. Thus, conductivity and resistivity need to be expressed as second-order tensors. In order to take the contribution of the magnetic field into account, Eq. 2.3 needs to be modified as

$$m^* \frac{d\boldsymbol{v}}{dt} = -\frac{m^* \boldsymbol{v}}{\tau} - e(\boldsymbol{E} + \boldsymbol{v} \times \boldsymbol{B}). \tag{2.6}$$

If the magnetic field is perpendicular to the 2DES ($\boldsymbol{B} = B\boldsymbol{e}_z$), the electric field generated by the current is

$$\begin{pmatrix} E_x \\ E_y \end{pmatrix} = \begin{pmatrix} \varrho_{xx} & \varrho_{xy} \\ \varrho_{yx} & \varrho_{yy} \end{pmatrix} \begin{pmatrix} j_x \\ j_y \end{pmatrix} = \begin{pmatrix} \frac{m^*}{n_e e^2 \tau} & \frac{B}{n_e e} \\ -\frac{B}{n_e e} & \frac{m^*}{n_e e^2 \tau} \end{pmatrix} \begin{pmatrix} j_x \\ j_y \end{pmatrix} =: \begin{pmatrix} \varrho_L & \varrho_H \\ -\varrho_H & \varrho_L \end{pmatrix} \begin{pmatrix} j_x \\ j_y \end{pmatrix}. \tag{2.7}$$

The equation above provides two important pieces of information:

- The longitudinal resistivity $\varrho_L$, which measures the resistivity along the direction of current flow, is equal to the zero field resistivity $\varrho_0$. Further, it is independent of the current direction ($\varrho_{xx} = \varrho_{yy} =: \varrho_L$) as expected for an isotropic system.
- An additional electric field builds up perpendicular to the current flow. This effect is called the Hall effect [19]. The electric field is proportional to the current with the proportionality constant

$$\varrho_H = \frac{B}{n_e e}. \tag{2.8}$$

Thus, the Hall effect provides a convenient method to determine either the magnetic field for a known charge density or the charge density when knowing the magnetic field strength.

The conductivity tensor is the inverse of the resistivity tensor and can be deduced from Eq. 2.7:

$$\sigma = \frac{1}{\varrho_L^2 + \varrho_H^2} \begin{pmatrix} \varrho_L & -\varrho_H \\ \varrho_H & \varrho_L \end{pmatrix} = \frac{1}{\left(\frac{m^*}{n_e e^2 \tau}\right)^2 + \left(\frac{B}{n_e e}\right)^2} \begin{pmatrix} \frac{m^*}{n_e e^2 \tau} & -\frac{B}{n_e e} \\ \frac{B}{n_e e} & \frac{m^*}{n_e e^2 \tau} \end{pmatrix} =: \begin{pmatrix} \sigma_L & -\sigma_H \\ \sigma_H & \sigma_L \end{pmatrix} \tag{2.9}$$

The equation above implicitly defines the following relations:

$$\sigma_L = \frac{\varrho_L}{\varrho_L^2 + \varrho_H^2} \tag{2.10}$$

$$\sigma_H = \frac{\varrho_H}{\varrho_L^2 + \varrho_H^2} \tag{2.11}$$

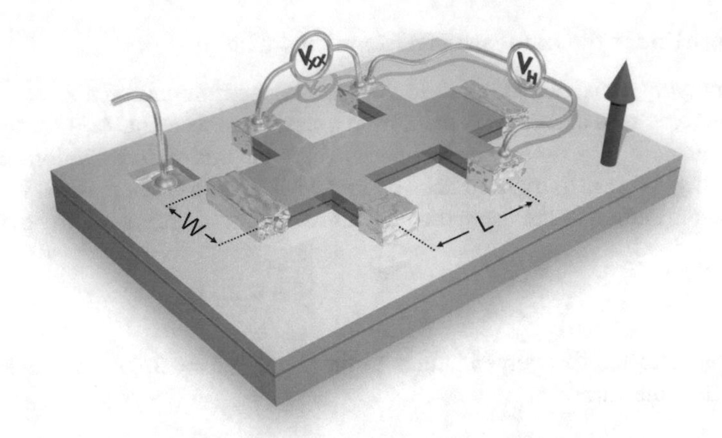

**Fig. 2.5** Schematic of a Hall bar structure used for most of the transport experiments in this thesis. The current is imposed through the front contact covering the entire width of the Hall bar. The longitudinal resistance is measured on neighboring contacts along the perimeter of the Hall bar, the Hall resistance between contacts on opposite sides. Typical dimensions are $W = 0.4\,\text{mm}$ and $L = 1.25$ mm. In many cases a backgate is present, which is separately contacted. The magnetic field (*brown arrow*) is directed perpendicular to the 2DES

In order to relate the resistivity to the quantity measured in experiment, i.e. the resistance $R$, the sample geometry and current flow must be well defined. Often a Hall bar structure as shown in Fig. 2.5 is used. Here, $W$ denotes the width of the Hall bar and $L$ the distance between two neighboring contacts. For this configuration, the longitudinal and Hall resistivity can be calculated as $\varrho_{xx} = R_{xx} \cdot \frac{W}{L}$ and $\varrho_H = R_H$ provided that the current flow is homogeneous. The samples used in this thesis typically have the dimensions $W = 0.4\,\text{mm}$ and $L = 1.25\,\text{mm}$.

## 2.3 The Integer Quantum Hall Effect

The Drude model introduced in the previous section predicts the Hall resistance to increase linearly with the magnetic field, while the longitudinal resistance is supposed to remain constant. This prediction is contrasted by the measurement of $R_{xx}$ and $R_H$ shown in Fig. 2.6.[2] In this measurement, a perpendicular magnetic field was applied to a high-quality sample at a temperature of roughly 20 mK. Already at a relatively low magnetic field of 100 mT, the Hall resistance deviates from its linear dependence, and $R_{xx}$ starts to oscillate. These deviations from the predicted behavior get stronger when ramping up the magnetic field. Eventually, $R_{xx}$ drops to zero in certain ranges of the magnetic field, interrupted by peaks of a finite resistance. In these regions

[2]Here, as in all other resistance measurements throughout this thesis, a low-frequency lock-in technique was used to improve the signal-to-noise ratio.

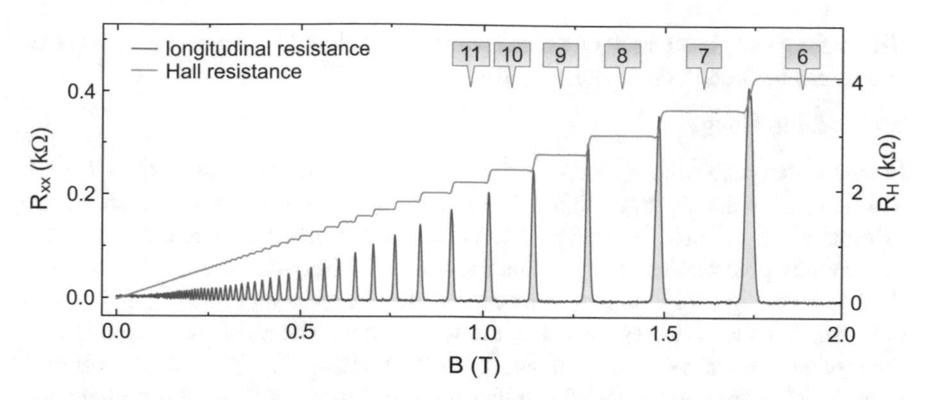


**Fig. 2.6** Longitudinal and Hall resistance for magnetic fields up to 2 T. Characteristic of the integer quantum Hall effect is a vanishingly small longitudinal resistance accompanied by a plateau in the Hall resistance. The integer values of $\nu$ in expression 2.12 are indicated in *blue* boxes

of $R_{xx} \approx 0$, the Hall resistance exhibits a plateau. The resistance values of these plateaus are precisely quantized to

$$R_H = \frac{h}{e^2} \cdot \frac{1}{\nu}, \tag{2.12}$$

with $\nu$ being an integer number ($\nu = 1, 2, 3, ...$). In view of these findings, this effect has been named the integer quantum Hall effect (IQHE). For its discovery, Klaus von Klitzing was awarded the Nobel Prize in 1985. The resistance values of the plateaus only depend on natural constants and are in particular independent of sample properties like charge density and spatial dimensions. For this reason, the IQHE is used as a resistance standard.

### 2.3.1 *Quantum Mechanical Treatment in the Single-Particle Picture*

The time-independent Schrödinger equation of non-interacting electrons in a magnetic field can be written as

$$\hat{H} \cdot \psi(\boldsymbol{r}) = \left[\frac{1}{2m^*}(\hat{\boldsymbol{p}} - e\boldsymbol{A}(\hat{\boldsymbol{r}}))^2 + e\phi(\hat{\boldsymbol{r}})\right]\psi(\boldsymbol{r}) = E \cdot \psi(\boldsymbol{r}). \tag{2.13}$$

The Hamilton operator $\hat{H}$ contains the momentum operator $\hat{\boldsymbol{p}}$, the electrostatic potential $\phi$ and the vector potential $\boldsymbol{A}$, which depends on the location operator $\hat{\boldsymbol{r}}$. $\psi$ are the eigenfunctions of the Hamilton operator. To calculate the solutions of the Schrödinger equation in a homogeneous magnetic field, two different gauges for the vector potential are frequently used, both of which are outlined below. The corresponding calculations can be found in textbooks [1, 20].

### The Landau Gauge

In the Landau gauge, the vector potential has only one component $\boldsymbol{A} = (0, -Bx, 0)$. This simplifies the algebra notably. In the case of free electrons, $\phi$ vanishes if no external electrical field is applied. For electrons confined to a quantum well, the electrostatic potential varies only along the growth direction ($z$-axis). In this case, the motion along the $z$-direction can be separated from the main Schrödinger equation, which leads to expression 2.1 in the first section. The normalized solution for the motion in the x–y–plane is shown in Eq. 2.14. Here, $H_n$ represents the Hermite polynomial of order $n$. Figure 2.7 displays the wavefunctions $\psi_{n,m}$ for different parameters $n$ and $m$. The eigenfunctions in Landau gauge comprise plane waves along the $y$-direction with wavevector $k_m$. For periodic boundary conditions, $k_m$ must fulfill $k_m = \frac{2\pi}{L_y} m$ with $m$ being an integer number ($m = 1, 2, 3...$) and $L_y$ representing the sample size in the $y$-direction. The plane waves are distributed equally along the $x$-axis and are centered at positions $x_m = l_B^2 k_m$.

$$\psi_{n,m}(x, y) = \frac{1}{\sqrt{2^n n! \sqrt{\pi} l_B}} H_n \left( \frac{x - l_B^2 k_m}{l_B} \right) e^{-\frac{\left(x - l_B^2 k_m\right)^2}{2 l_B^2}} e^{i k_m y} \tag{2.14}$$

The solutions of the Schrödinger equation contain the magnetic length $l_B = \sqrt{\hbar/(eB)} \approx 25.7/\sqrt{B[T]}$ nm—a characteristic length scale brought about by the magnetic field. The physical consequences of the eigenfunctions are discussed below. They must be independent of the chosen gauge. Properties of the wavefunctions that do depend on the gauge, like the orientation of the plane waves along the $y$-direction, have no physical meaning. In fact, one can also choose a gauge which is characterized by the circular symmetry of its eigenfunctions as described in the following.

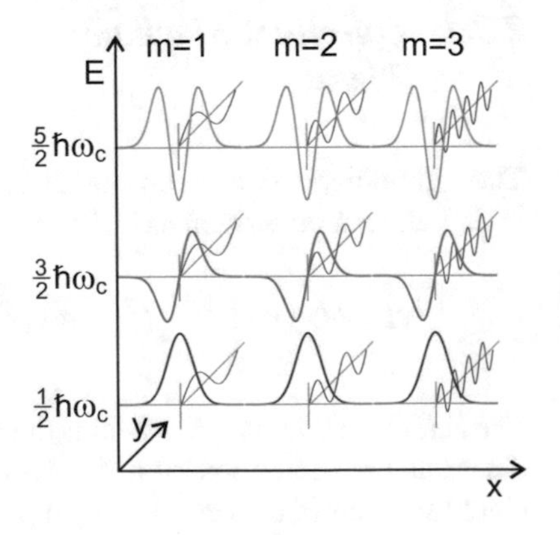


**Fig. 2.7** Electron wavefunctions in the Landau gauge for different parameters of $m$ and $n$ (based on [3]). Both energy and wavefunction amplitude are plotted along the $z$-axis

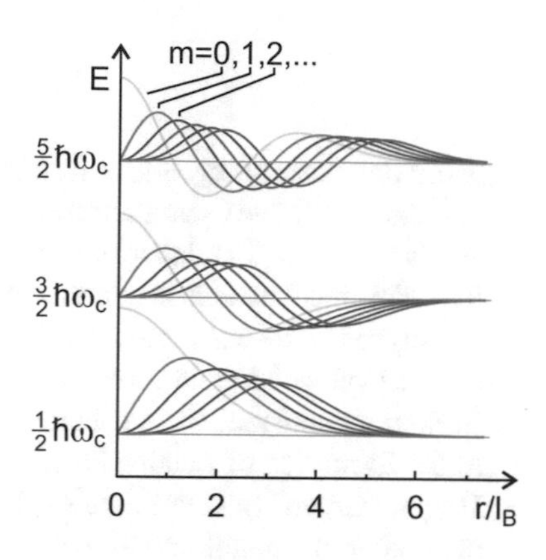


**Fig. 2.8** Electron wavefunctions in the symmetric gauge for different parameters of $m$ and $n$ (based on [3]). Both energy and wavefunction amplitude are plotted along the ordinate

### The Symmetric Gauge

The symmetric gauge is rotationally invariant about the $z$-axis and is defined as $\boldsymbol{A} = \frac{1}{2}\boldsymbol{B} \times \boldsymbol{r}$. Consequently, the eigenfunctions in the symmetric gauge are best parameterized in polar coordinates $r$ and $\theta$:

$$\psi_{n_r,m}(r,\phi) = \frac{1}{l}\sqrt{\frac{n_r!}{(n_r+|m|)!}}\left(\frac{r}{l_B\sqrt{2}}\right)^{|m|} e^{-\frac{r^2}{4l_B^2}} L_{n_r}^{|m|}\left(\frac{r^2}{2l_B^2}\right) e^{im\theta}, \tag{2.15}$$

where $L_{n_r}^{|m|}$ are the associated Laguerre polynomials. The wavefunctions are shown graphically in Fig. 2.8. The radial quantum number $n_r$ ($n_r \geq 0$) counts the zeros in the radial term of the wavefunction and relates to $n$ (in the Landau gauge) via $n = n_r + (|m| - m)/2$. The angular momentum quantum number $m$ is restricted to $m \geq -n$. For large values of $m$, the shape of the wavcfunctions in the symmetric gauge is roughly equal to rings of width $l_B$ and radius $r_{|m|} = \sqrt{2|m|}l_B$. As in the case of the Landau gauge, the wavefunctions are shaped along the vector potential $\boldsymbol{A}$. Because of the centrifugal potential, the wavefunctions get pushed away from the origin when increasing $m$. In general, the solutions in the symmetric gauge are more complicated to handle than those in the Landau gauge. Therefore, preferably the Landau gauge is used, except for cases where rotational symmetry is prevalent, e.g. in quantum dots.[3]

### Gauge Independent Properties

The following consequences of the Schrödinger equation are independent of the gauge choice and describe important properties of the 2DES:

- **Formation of Landau levels**—The eigenvalues of the wavefunctions yield the energy spectrum of electrons in a magnetic field. They can be expressed as

[3]In some cases, which are not of relevance for this work, it is advantageous to use the von Neumann lattice gauge. Here, the wavefunctions describe a cyclotron motion of electrons centered around points of a von Neumann lattice [21].

$$E_n = \left(n + \frac{1}{2}\right) \cdot \hbar\omega_c, \tag{2.16}$$

using the classical cyclotron frequency $\omega_c = eB/m^*$. Thus, the energy spectrum has discrete values parameterized by the index $n$ ($n = 0, 1, 2...$). These energy levels are called Landau levels. They are separated by the cyclotron energy $E_c = \hbar\omega_c$. This result stands in contrast to classical calculations which predict a constant density of states (Eq. 2.2). The reduction of the electronic energy spectrum to discrete values brought about by the magnetic field has far-reaching consequences. In fact, the existence and robustness of the quantum Hall effect is deeply rooted in the formation of Landau levels.

- **Degeneracy of Landau levels**—The fact that the energy values in Eq. 2.16 only depend on the quantum number $n$ and not on $k$ (in Landau gauge) shows that the Landau levels are highly degenerate. The number of allowed states per unit area in a Landau level $n_L$ depends on the sample dimensions. In the Landau gauge, we need to impose the requirement that the center coordinate $x_m$ lies within the sample, i.e. $x_m = l_B^2 \frac{2\pi}{L_y} m < L_x$. This immediately leads to the expression

$$n_L = \frac{eB}{h} = \frac{B}{\Phi}. \tag{2.17}$$

Thus, a Landau level can accommodate as many electrons as magnetic flux quanta $\Phi = h/e$ penetrate the sample. In a sense, each electron state occupies the area of a single flux quantum. An important quantity in the context of the quantum Hall effect is the filling factor $\nu$. It denotes how many Landau levels are filled with electrons. Using Eq. 2.17, the filling factor can be calculated as

$$\nu = \frac{n_e}{n_L} = \frac{n_e h}{eB}. \tag{2.18}$$

Strictly speaking, this expression is only valid on length scales much larger than the extent of the electron wavefunction. For small dimensions, the electron density $n_e$ in Eq. 2.18 has to be replaced by the density of the wavefunctions' center coordinates. Of course, at macroscopic dimensions both definitions are equivalent.

### The Spin Degree of Freedom

Up to now, the implications of the electron spin have been swept under the rug. If an electron spin is placed in a magnetic field, it contributes an additional energy $E_z$ alongside the cyclotron energy $E_c$ due to the Zeeman effect. The Zeeman energy depends linearly on the magnetic field:

$$E_z = g_e^* \mu_B B, \tag{2.19}$$

with $\mu_B = e\hbar/2m$ being the Bohr magneton and $g_e^*$ the effective g-factor. For electrons in bulk GaAs, $g_e^*$ has a value of $-0.44$ [22, 23]. However, in a 2DES, exchange interaction effects might greatly modify $g_e^*$ at low filling factors [23]. Equation 2.19 implicitly assumes that conduction band electrons in GaAs have total spin 1/2 (s-band, orbital spin 0). Hence, the Zeeman effect splits each Landau level into a spin-up and spin-down branch separated by $E_z$. The additional degree of freedom brought by the electron spin doubles the density of states in a Landau level (factor 2 in Eq. 2.17). Nevertheless, it is customary for the definition of the filling factor (Eq. 2.18) to count the spin branches as separate levels: $\nu = 1 \rightarrow$only the spin-up branch is filled $(n = 0, \uparrow)$; $\nu = 2 \rightarrow$the spin-down branch $(n = 0, \downarrow)$ is filled as well.

Figure 2.9 shows schematically the Landau level structure as a function of magnetic field. The energy spectrum has a fan-like structure since both Zeeman and cyclotron energy increase linearly with the magnetic field. The relative weight of the respective energies is understated in Fig. 2.9. For typical experimental conditions, the cyclotron energy is about 70 times larger than the Zeeman energy (for $g_e^* = -0.44$ and $m^* = 0.067\,m$). This pronounced difference is the reason why integer quantum Hall states (IQHS) are generally more stable at even filling factors, i.e. they have a higher energy gap compared to odd filling factors (even in the presence of exchange enhancement at lower filling factors). Figure 2.9 also shows the evolution of the Fermi energy when increasing the magnetic field for a fixed electron density. $E_F$ follows the linear increase of the Landau levels up to the point where the degeneracy in the lower levels is high enough to accommodate all electrons of the upper level. At this point, the Fermi energy jumps to the lower level, which creates the sawtooth-like

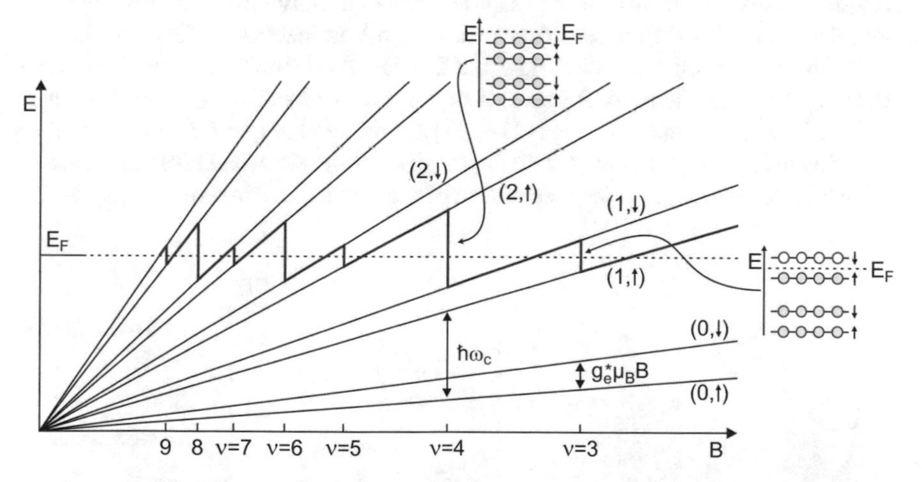


**Fig. 2.9** Evolution of the discrete Landau level energies in a perpendicular magnetic field (based on [3]). The Landau levels are separated by the cyclotron energy. In addition, each Landau level is split by the Zeeman energy (splitting is exaggerated for clarity). Both energies depend linearly on the magnetic field. In addition, the Landau level degeneracy increases when raising $B$, and more electrons can be accommodated in the lower-lying Landau levels. Once a level is emptied completely, the Fermi energy (*red line*) drops abruptly

pattern shown in Fig. 2.9. Since the Fermi energy jumps between levels of opposite spin orientation, the total spin polarization exhibits an oscillatory behavior. At even filling factors, the electron system is unpolarized and becomes increasingly polarized towards (smaller) odd filling factors. Deviations from this general behavior arise from many-particle interactions between the electrons. We will return to this important point in later chapters.

## 2.3.2 *Microscopic Picture*

Many aspects of the QHE cannot be understood without looking at the electrostatic situation of the sample at a microscopic level. For instance, the width of the resistance plateaus in Fig. 2.6 depends decisively on the disorder potential experienced by the electrons, and the electrostatic potential at the edges strongly influences the current flow through the sample.

### The Role of Disorder

The Landau levels introduced above as a solution of the Schrödinger equation have a sharp, peak-like density of states. In reality, the disorder present in the sample due to background impurities and remote ionized donors leads to an energetic broadening $\Gamma$ of the Landau levels. The exact shape of the broadening is unclear and presumably depends on the details of the disorder potential. Some publications support a Gaussian shape [24–26], while others reported a Lorentzian one [27, 28]. Uncertainty governs also the question how the Landau level broadening evolves when changing the magnetic field. A square root dependence has been observed in references [25, 26], whereas references [24, 27] found no change of $\Gamma$ with magnetic field. Figure 2.10a depicts the $B$-field dependence of the electron DOS in the case of a Gaussian broadening. A clear separation of the Landau levels requires $\hbar\omega_c > \Gamma$.

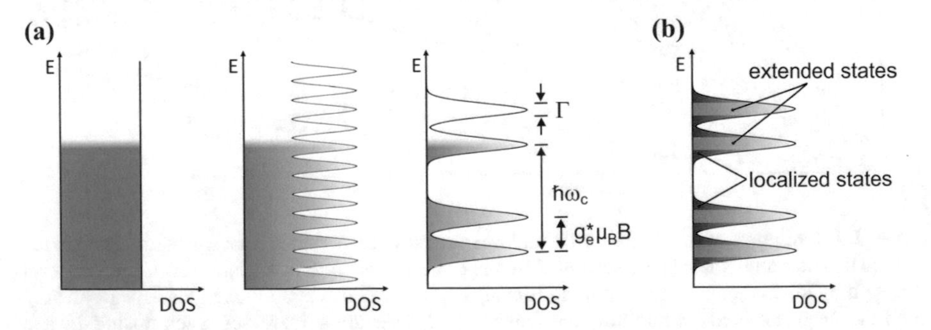


**Fig. 2.10** **a** Evolution of the electron DOS with increasing magnetic field. Landau levels develop separated by the cyclotron energy. The Zeeman energy splits each Landau level into two branches of opposite spin orientation. **b** Due to the disorder in the sample, each Landau level branch is divided into extended states at its center and localized states at its periphery

The disorder potential not only leads to a broadening of the Landau levels but also divides the electron system into localized and extended states (Fig. 2.10b). Intuitively, electrons are localized either in the valleys or hills of the disorder landscape. The motion of the confined electrons can in a simplified picture be viewed as closed orbits of constant energy inside of the hills and valleys as shown in Fig. 2.11. Quantum mechanics dictates that these orbits enclose an integer number of flux quanta [2]. The localized states become populated predominantly for high and low fillings of a Landau level. For intermediate filling factors, the electrons are free to move across the sample. Yet, this simplified picture does not grasp the whole complexity of the problem. Ilani et al. found the number of localized states to be independent of the magnetic field by probing the 2DES with a single-electron transistor [29]. This is surprising in view of the magnetic field dependent Landau level degeneracy. As the magnetic field and with it the flux density increases, more and more localized states are expected to form in the valleys and hills of the disorder landscape [29]. Thus, in the single-particle picture, the density of localized states would depend on the magnetic field. Based on this discrepancy, the authors concluded that the localization physics in high-quality (low disorder) samples is dominated by Coulomb interaction. Further, each of the electron pockets has to be treated as a quantum dot (or anti-dot). Because of the Coulomb repulsion inside of the quantum dot (Coulomb blockade), each electron pocket can accommodate only a fixed number of electrons. As a consequence, the density of localized states is independent of the magnetic field.

Given the separation of the DOS into localized and extended states, it is possible to understand intuitively the vanishing of the longitudinal resistance over a wide magnetic field range in the IQHE (Fig. 2.6). Whenever the Fermi energy falls into a region of localized states, the conductivity $\sigma_{xx}$ drops to zero, and according to Eq. 2.10, $\varrho_{xx} = 0$ follows. If extended states are present at the Fermi energy, $R_{xx}$ has a finite value. The large number of localized states ensures $R_{xx} \approx 0$ over an extended filling factor range in contrast to the situation in Fig. 2.9, where zero longitudinal

**Fig. 2.11** Visualization of localized states in the single-particle picture. Shown is the spatial variation of the electrostatic potential in a disordered sample. Electrons get trapped in the valleys and hills of the disorder landscape. Their orbits are indicated in *blue*

resistance would only be observable at the point when $E_F$ jumps from one Landau level to the other.

What we have not addressed so far is the question why the Hall resistance is quantized. To answer this question, we have to take a closer look at the edges of the sample.

### The Landauer-Büttiker Picture

When going away from the interior of the sample towards the edges, the Landau levels have to bend upwards because the Fermi energy at the surface is pinned to a value within the GaAs bandgap (Fig. 2.12a). Consequently, the Landau levels coincide with the Fermi energy at the edges, and one-dimensional conductive channels are created, even if the interior part of the sample is insulating. For each completely filled Landau level branch, one transport channel is created on either side of the sample (Fig. 2.12b). The electron transport along the two sides occurs in opposite directions as determined by the potential gradient at the edges. The spatial separation of the counter-propagating current paths strongly suppresses dissipative backscattering of the charge carriers. As a consequence, a constant potential is maintained throughout each channel, and the voltage drop between contacts on the same side becomes vanishingly small. At the same time, each one-dimensional channel adds a value of $e^2/h$ to the total conductance [30]. Based on this argument, it becomes obvious that the Hall resistance is precisely quantized to $R_H = h/(e^2\nu)$ simply because the Hall voltage is set by the potential difference between opposite sides of the sample. This approach of decomposing the transport in the quantum Hall regime into isolated current channels goes back to ideas by R. Landauer and M. Büttiker [30, 31].

### The Interacting-Electron Picture

When taking a closer look, the Landauer-Büttiker picture of current flow has some obvious shortcomings. For instance, the sudden step-like drop of the electron density

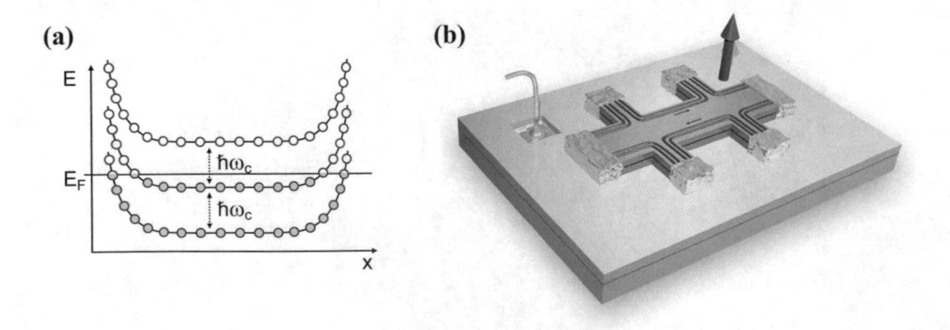


**Fig. 2.12** The Landauer-Büttiker edge channel picture. **a** Spatial evolution of the Landau level energy across the sample. At the edges, the Landau levels are lifted by the confinement potential. As a consequence, partially filled one-dimensional channels are created which run along the sample edges. **b** Visualization of the edge channels in a Hall bar structure

at the edges of the sample is energetically unfavorable (Fig. 2.13a left). In a more realistic scenario, the Coulomb repulsion between electrons would lead to a smoothly varying density decrease towards the edges. This is confirmed by the self-consistent calculations of the electrostatic and chemical potential performed by Chklovskii et al. [32, 33]. Their findings are shown in Fig. 2.13a (right). In contrast to the single-particle picture of Landauer and Büttiker, the self-consistent calculations separate the sample into regions of opposing character—compressible and incompressible stripes. In the *compressible* stripes, the topmost filled Landau level is only partially occupied, and its energy is fixed to the Fermi energy. In the *incompressible* stripes, all occupied Landau levels are completely filled, and the electrons cannot screen the lateral confinement potential. Hence, the absolute value of the chemical potential decreases here towards the edge. All incompressible stripes have an integer value of the filling factor. Scattering of the electrons inside of these stripes is therefore strongly suppressed by the energy gap to the next, unoccupied level. As a result, the current can flow dissipationless inside of the incompressible stripes, and any externally injected current will preferentially distribute to flow in the incompressible stripes. More precisely, the current will favor the innermost incompressible stripe on either side of the sample because it is the most stable one and can sustain the highest currents [33–35]. If these incompressible stripes become too narrow (on the order of $l_B$), scattering into the central compressible regions is possible, and dissipation occurs. Hence, a quantized Hall plateau is only observable if at least one incompressible stripe exists in the sample which is wide enough to sustain a dissipationless current.

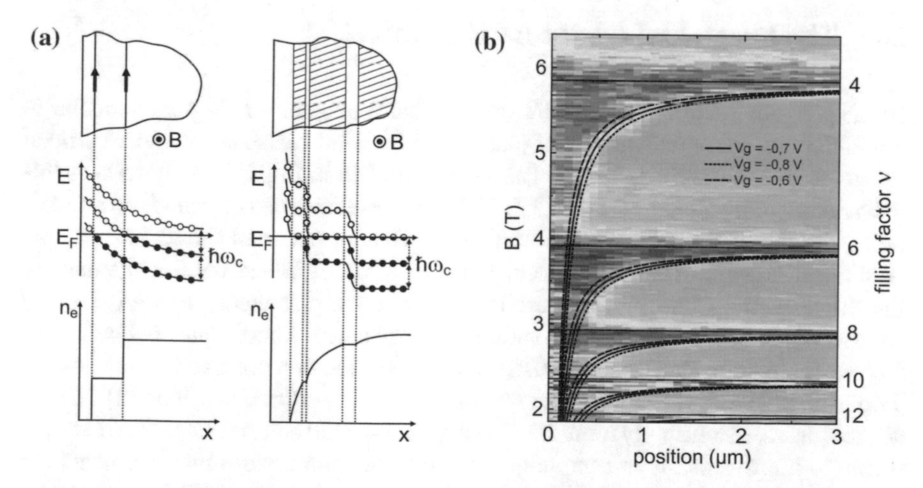


**Fig. 2.13** The interacting-electron picture. **a** The Coulomb interaction favors a smooth density distribution at the edges of the sample (*right*) in contrast to the Landauer-Büttiker picture (*left*). This leads to the formation of compressible (hatched regions) and incompressible stripes (picture modified from [32]). **b** Scanning probe measurement of the Hall potential distribution from the edge towards the center of the sample (picture taken from [40]). Black lines indicate the theoretical position of the innermost incompressible stripe (see [40] for details)

The behavior described above has been confirmed experimentally by measuring the Hall potential distribution across the sample using different scanning probe techniques with either a metal tip [36, 37] or a single-electron transistor [38, 39] as a sensor. The spatial variation of the Hall potential provides information about the current distribution because the Hall potential drops in a region where the current flows. An example of a measured Hall potential distribution is depicted in Fig. 2.13b. It shows that the current distribution moves to the interior part of the sample as the magnetic field approaches an integer value of the filling factor. This observation is consistent with the theoretical movement of the innermost incompressible stripe (black lines). Details on these experiments are given in reference [40].

Yet, the question remains how the quantization of the Hall resistance comes about in this model. The very fact that in the quantum Hall effect the current flows in the innermost incompressible stripe is the reason why the Hall resistance is precisely quantized to $R_H = h/(e^2\nu)$. This explanation is straightforward when bringing to mind that the innermost incompressible stripe always has an integer filling factor and its value is the largest integer filling factor in the sample. If the current flow is limited to this filling factor region, the Hall voltage drop will be as well. Combining Eqs. 2.8 and 2.18 then directly leads to $R_H = h/(e^2\nu)$. The onset of the quantized resistance plateau therefore occurs at the point when the innermost incompressible stripe is wide enough to maintain its incompressible character. By increasing the magnetic field, this stripe will move to the center of the sample. Eventually, the plateau will disappear once the central stripe gets disrupted by disorder and vanishes [35].

## 2.4 The Fractional Quantum Hall Effect

In the previous section, we have introduced that a magnetic field perpendicular to the 2DES can cause a sequence of plateaus in the Hall resistance centered around integer values of the filling factor. The sample studied in Fig. 2.6 nicely follows this behavior up to a magnetic field of 3 T. If we increase the magnetic field beyond this point, deviations from the simple oscillatory behavior occur as shown in Fig. 2.14. Additional quantum Hall states occur in between the IQHS at fractional values of the filling factor. Clearly visible are for example the plateaus at $\nu = 5/3$, $4/3$ and $4/5$ accompanied by a vanishing longitudinal resistance. These states belong to the fractional quantum Hall effect (FQHE). The FQHE was first observed for $\nu = 1/3$ by Tsui et al. in 1982 [41] shortly after the discovery of the IQHE in 1980 [42]. In subsequent years, increasingly more fractional quantum Hall states (FQHS) were found, owing to improvements in sample quality and low-temperature measurements. At the present day, more than 80 different FQHS are known to exist [43]. The discovery and explanation of the FQHE has been honored with awarding the Nobel Prize to Robert Laughlin, Horst Störmer and Daniel Tsui in 1998.

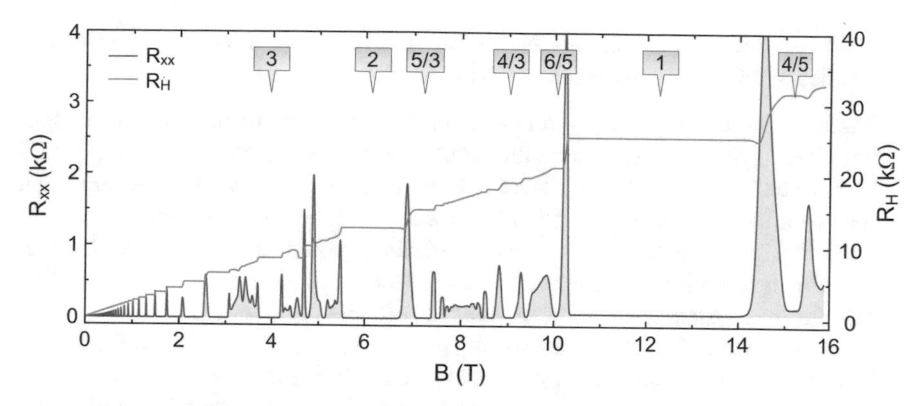


**Fig. 2.14** Longitudinal and Hall resistance measured at base temperature (~20 mK) over a broad magnetic field range. Multiple fractional quantum Hall states are visible in between the IQHS. Filling factors are indicated in *blue* boxes

The existence of quantum Hall states at fractional filling factors came as a surprise because it cannot be explained by the single-particle theory introduced in the previous section. At a partial filling of Landau levels, the electron system is supposed to be compressible, and no excitation gap for the formation of quantum Hall states should exist. The FQHE must therefore originate from electron-electron interactions. In partially filled Landau levels, electrons can rearrange and thereby reduce their Coulomb energy while maintaining a constant kinetic energy. This is not possible for the IQHE because of the completely filled Landau levels. For this reason, the Coulomb interaction does not enter the IQHE in the first place. Yet, when taking a closer look, the Coulomb interaction can also significantly affect the IQHE as pointed out in the previous section. Describing these many-particle interactions theoretically is in general an exceptionally challenging endeavor. Nevertheless, several ingenious theoretical concepts have been put forward [2, 44–49]. The two most important ones are introduced hereinafter.

### 2.4.1 *Laughlin's Wavefunction*

In an attempt to describe the formation of quantum Hall states at fractional filling factors $\nu = {}^1\!/_q$ in the lowest Landau level, Laughlin proposed a succinct trial wavefunction

$$\psi_{1/q} = \prod_{i<j} (z_i - z_j)^q \exp\left( -\frac{1}{4l_B^2} \sum_i |z_i|^2 \right), \tag{2.20}$$

where $z_i$ denotes the position of the $i$th electron in complex numbers [44, 50]. The properties of this wavefunction are described below:

- Based on the assumption that all electrons in the lowest Landau levels are spin polarized (symmetric spin wavefunction), the orbital wavefunction in Eq. 2.20 must be anti-symmetric. This forces $q$ to be an odd integer number. Assuming a spin-polarized electron system is in fact only justified for filling factors $^1/_q$. At other FQHS the electron system can be partially polarized or even unpolarized. We will come back to this important issue at a later point.
- The electron system has a homogeneous density distribution [2].
- The second part of Eq. 2.20 consists of a product of Gaussian wavefunctions reminiscent of the single-particle solutions discussed in the previous section.
- The product term at the beginning contains the electron-electron interactions by imposing a $q$-fold zero point whenever two electrons are at the same position. This ensures that all electrons are well separated and thereby reduces the Coulomb energy. In a more advanced interpretation, the zeros in the wavefunction can be understood as vortices, i.e. as points of vanishing probability density in the center and a phase change of $2\pi$ when circulating around the vortex. Intuitively, one can imagine each electron to carry $q$ vortices with it. The correlation of electrons and vortices ensures that no two electrons can occur in the same place. One vortex would be enough to fulfill Pauli's exclusion principle. However, the additional $q-1$ vortices further reduce the electrostatic Coulomb energy considerably [51]. This is the basic idea behind the electron-electron correlations in a 2DES exposed to a strong magnetic field. Besides the $q$ vortices at each electron's location, no further vortices are present in the 2DES. Hence, the number of vortices equals the number of magnetic flux quanta piercing through the sample simply because of the respective filling factor $1/q$. In fact, one can think of each vortex in the 2DES as embodiment of a single flux quantum [52]. This close relation between flux quanta and vortices as well as their affinity to electrons forms the basis of the composite fermion theory detailed below.

Above, we have introduced that a 2DES at $\nu = {}^1/_q$ can minimize its Coulomb energy by attaching $q$ vortices to each electron. When moving away from this commensurate electron-vortex ratio, e.g. by increasing the magnetic field and adding another vortex, a substantial amount of Coulomb energy has to be paid [46, 51]. This determines the energy gap of the FQHE and therby identifies the Coulomb energy as the prevalent energy scale for FQHE physics. At the same time, introducing additional vortices in the 2DES excites quasiparticles that come along with very surprising properties. Because of the strong tie between vortices and electrons in Laughlin's theory, introducing one additional flux quantum can be understood equivalently as adding a quasiparticle with a fractional charge $e^* = e/q$ [44]. This startling property of Laughlin's quasiparticles has far-reaching consequences as these particles obey neither Fermi-Dirac nor Bose-Einstein statistics [46]. Instead, they follow a more general anyonic statistics, which will be addressed in Sect. 2.6.

Yet, the question remains whether Laughlin's wavefunction is indeed correctly describing FQHE physics and, if so, to what extent. The existence of fractionally charged excitations has been confirmed experimentally by tunneling [53] and shot noise measurements [54–56] as well as more recently in experiments using single-electron transistors [57]. The verification of these counterintuitive entities provides strong support for Laughlin's theory. Apart from that, it is a beautiful example of fractionalization in solid state physics, according to which the properties of quasiparticles cannot be understood simply as a linear combination of its elementary constituents. This general phenomenon has been encountered in different manifestations in recent years, for example as magnetic monopoles in spin ice materials [58–60] or as the deconfinement of an electron into a holon, spinon and orbiton [61–63]. Further support for Laughlin's theory comes from exact diagonalization studies [64] and Monte Carlo simulations [65], where it was shown that Laughlin's wavefunction yields the correct ground state for short range interactions.

Laughlin's concept has been successfully generalized to FQHS at filling factors different from $1/q$. Quantum Hall states at filling factors $\nu = 1 - 1/q$ follow directly from particle-hole symmetry [51, 66]. Wavefunctions for other fractional filling factors were constructed by Halperin and Haldane using a hierarchical approach based on Laughlin's quasiparticles [45, 46, 51]. However, these wavefunctions are more complex and partly lead to discrepancies between experiment and theory, such as the relative strength of FQHS [2]. An alternative and at the same time very elegant approach to construct higher-order FQHS is the composite fermion model.

### 2.4.2 The Composite Fermion Theory

In 1989 Jainendra Jain enriched the understanding of the FQHE by proposing a new concept [67]. Again, it is based on the coupling of electrons and vortices, but this time an electron dressed with an even number of magnetic flux quanta is considered as a new entity—the composite fermion (CF). With this transformation in mind, the FQHE can be elegantly rephrased as the IQHE of composite fermions. In the following, we illustrate this statement at an intuitive level using Fig. 2.15. For a deeper understanding, we refer to reference [2]. Figure 2.15 shows the longitudinal resistance for the FQHS around $\nu = 1/2$ in the lower panel. Plotted above are the IQHS appearing at smaller magnetic fields. Filling factor $\nu = 1/2$ corresponds to the situation where twice as many magnetic flux quanta penetrate the sample as electrons are present. After attaching two flux quanta to each electron, the resulting composite particles essentially move around in a vanishing effective magnetic field because all flux quanta were already taken into account in the construction of the composite fermions. Hence, a compressible Fermi sea of composite fermions is formed. When increasing the magnetic field and going away from $\nu = 1/2$, the effective magnetic field $B^* = B - B(\nu = 1/2)$ increases likewise. As a consequence, quantum Hall states start to develop for integer values of the effective filling factor $\nu^* = n_e h/(eB^*)$. Figure 2.15 highlights the essence and beauty of the CF model: The IQHE of composite fermions in their effective magnetic field coincides with the FQHE of electrons in the total magnetic field. In this sense, the FQHE at

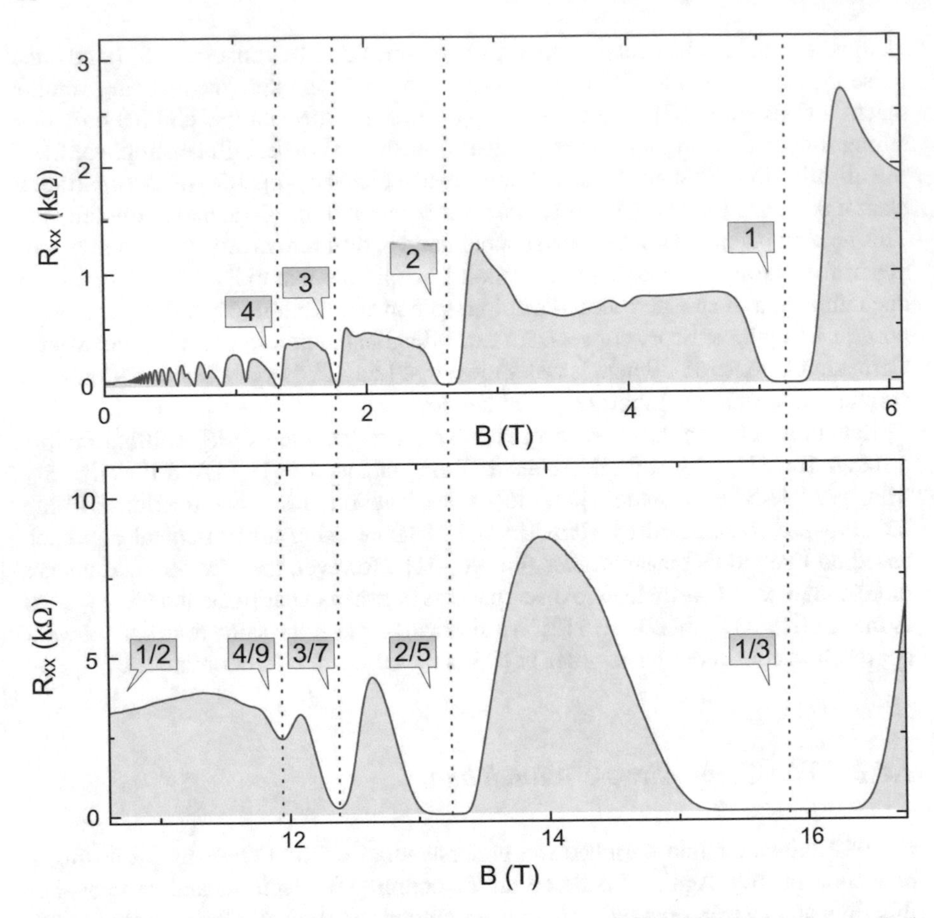


**Fig. 2.15** Visualization of the CF picture (plotted according to [78]). Shown is the longitudinal resistance up to $\nu = 1$ (*upper panel*) and between $\nu = 1/2$ and $\nu = 1/3$ (*lower panel*). The positions of the FQHS with respect to $\nu = 1/2$ coincide with the IQHS in the total magnetic field. The *upper panel* was measured at roughly 400 mK in order to locate more easily the position of each IQHS

$\nu = 1/3$ ($2/5$, $3/7$, ...) can be understood as the IQHE of composite fermions with $\nu^* = 1$ (2, 3, ...). Thus, the CF theory transforms the strongly interacting 2DES into a system of weakly interacting composite fermions whose energy spectrum develops discrete Landau levels if $B^*$ is strong enough. These CF Landau levels are sometimes referred to as $\Lambda$ levels. Drawing further on the analogy to the IQHE, the excitation gap of composite fermions can be interpreted as cyclotron energy $E_c = \hbar e B^*/m^*_{CF}$, with $m^*_{CF}$ being the effective CF mass. $m^*_{CF}$ must be proportional to $\sqrt{B}$ since the energy gap arises solely from Coulomb interactions.

In general, the CF theory accounts for FQHS at filling factors

$$\nu = \frac{\nu^*}{p \cdot \nu^* \pm 1}. \tag{2.21}$$

In this expression, $p$ is an even integer number and denotes the number of flux quanta attached to each electron. Equation 2.21 covers only filling factors in the lowest Landau level branch. However, it can be extended to the second Landau level by ignoring the completely filled lower-lying levels. This accounts for the FQHE at $\nu = {}^{7}/_{3}$ and $\nu = {}^{10}/_{3}$, for instance. In addition, particle-hole symmetry within each Landau level might be observable, leading to states such as $\nu = {}^{4}/_{5}$ and $\nu = {}^{5}/_{7}$.

The composite fermion theory has been proven very helpful in explaining various experimental results [68–70]. The validity of the CF concept is further supported by the determination of the CF Fermi wavevector [71, 72] and the demonstration of a quasi-classical motion of the composite fermions in their effective magnetic field $B^*$ [73–75]. Cyclotron resonances of composite fermions have been measured directly by microwave absorption [76]. Interestingly, also the FQHE of composite fermions has been observed, which highlights the existence of residual interactions between composite fermions [77].

## 2.5 Density-Modulated Phases in the Quantum Hall Regime

When looking at Fig. 2.14 in the previous section, it becomes apparent that the FQHS only appear at high magnetic fields in the first and second Landau level ($n = 0, 1$). At lower magnetic fields, i.e. when higher Landau levels are occupied, density-modulated phases are energetically favored at partial fillings [79]. Their existence is not directly visible in Fig. 2.14 but will be demonstrated experimentally later on. For these density-modulated phases, the homogeneous electron system in the topmost Landau level breaks up into alternating regions of high and low electron density. Two types of density-modulated phases are believed to exist [80]: the bubble phases characterized by a twofold periodicity and the stripe phases, which consists in the simplest case of parallel stripes with a strictly one-dimensional (1D) periodicity (Fig. 2.16). In both cases, the system alternates between regions where the topmost Landau level branch is completely filled and regions where it is completely empty. The reason for the formation of density-modulated phases is rooted in the competition between the repulsive direct Coulomb interaction on the one hand and the attractive exchange interaction on the other hand [80, 81]. Whenever the 2DES deviates from a homogeneous distribution, this comes at the cost of direct Coulomb energy. Despite the abrupt change of the filling factor in Fig. 2.16, the impact on the Coulomb interaction is weakened by the spatial extent of the electron wavefunction in the plane of the 2DES as it smoothens the electron density distribution. At dimensions on the order of the wavefunction width, it is important to bear in mind that the filling factor is given by the density of the wavefunctions' center coordinates and not by the electron density (see Sect. 2.3.1). If the modulation period is similar to the width of the wavefunction ($\sim l_B$), the density modulation and therefore also the increase in Coulomb energy will be small in comparison to the step-like change of the filling factor. The

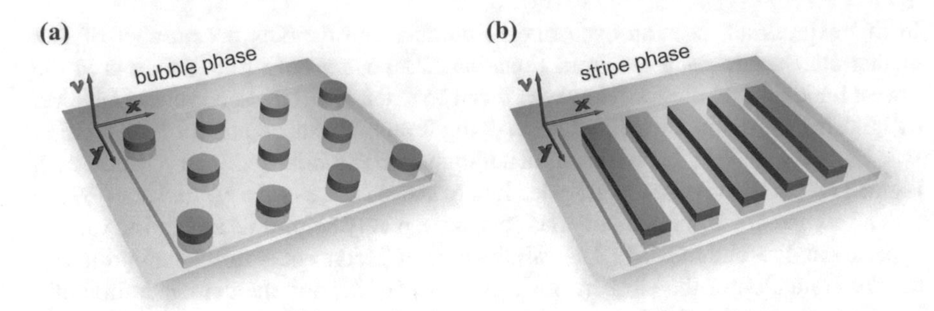


**Fig. 2.16** Spatial modulation of the filling factor in the topmost Landau level for the bubble and stripe phase. **a** In the case of the bubble phase, the homogeneous electron system contracts to islands which arrange in a triangular lattice. **b** The stripe phase may in the simplest case be understood as a periodic arrangement of parallel stripes with a strictly one-dimensional periodicity

gain in exchange energy, however, is rather determined by the modulation of the filling factor [82]. In total, the 2DES can minimize its energy by modulating the electron density at certain fillings of the topmost occupied spin branch [79, 82, 83]. The average filling factor determines the phase symmetry, i.e. whether the bubble or the stripe phase is favored energetically. In general, the energy gain of density-modulated phases is rather small. Hence, to observe such phases, low temperatures ($\sim$100 mK) as well as high mobilities ($\sim 10^7$ cm$^2$/Vs) are required.

### *2.5.1 The Bubble Phase*

The bubble phase is present in Fig. 2.14 though not yet visible. Its presence becomes evident as one slightly raises the temperature to about 35 mK. The result is shown in Fig. 2.17. At higher temperatures, the physics appears to be much richer than the low-temperature data suggests. The blue curve reveals that the IQHS breaks down rather quickly when increasing (and lowering) the magnetic field. However, the IQHE reappears at even higher (lower) magnetic fields, where $R_{xx}$ drops to zero and $R_H$ returns to the value of the nearby IQHE plateau. These states are therefore called reentrant integer quantum Hall states (RIQHS). In the case of $n \geq 2$, the RIQHS occur at elevated temperatures around the partial filling factors $\tilde{\nu} = 1/4$ and $\tilde{\nu} = 3/4$, where $\tilde{\nu}$ denotes the filling factor of the highest occupied Landau level branch [84–86]. If the electron temperature is low enough, a seamless transition to the IQHE is observed in Fig. 2.17. The reappearance of the IQHE can be understood as a manifestation of the bubble phase: Since the lattice of electron clusters is pinned by disorder, it cannot participate in transport. Therefore, the system behaves in a standard charge transport experiment as if all Landau levels are filled, and the IQHE occurs. The same arguments hold for the RIQHS at $\tilde{\nu} = 3/4$ with the exception that there clusters of holes are embedded in the electron system of the next higher integer filling factor.

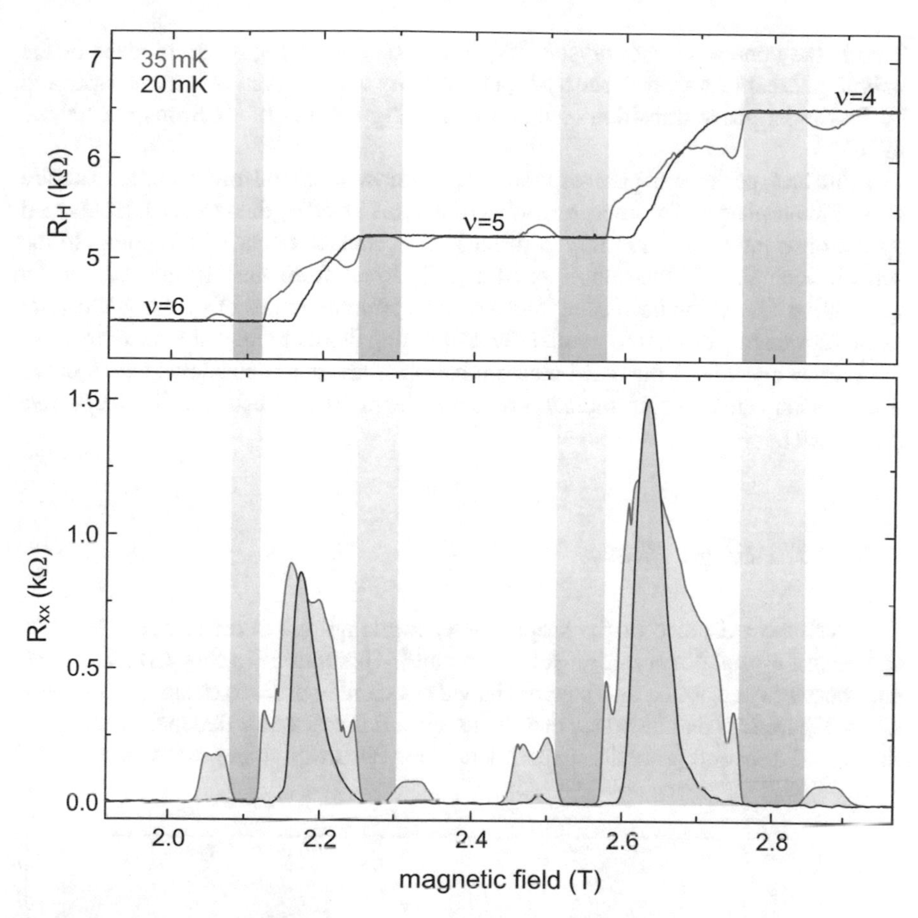


**Fig. 2.17** Longitudinal and Hall resistance at base temperature (~20 mK, in *red*) and at slightly higher temperatures (~35 mK, in *blue*). At elevated temperatures a reappearance of the IQHE is visible (marked in *green*)

Extensive theoretical studies of the bubble phase and, in general, the density-modulated phases have been carried out. They predict that the electron bubbles arrange in a triangular lattice with a period of roughly $3 \cdot R_c$, where $R_c = \sqrt{2\pi n_e}\hbar/(eB)$ denotes the cyclotron radius [79, 82]. The number of electrons varies according to the partial filling $\tilde{\nu}$ with the restriction that the maximum number of electrons per bubble is limited to $n$ in the $n$th Landau level (for $n > 1$) [79, 87].[4] The optimal number of electrons per bubble can be well approximated by $3\tilde{\nu}n$ [80, 87]. Thus, when increasing $\tilde{\nu}$ from $\tilde{\nu} = 0$, at first, only a single electron occupies a bubble site.

[4]Côté et al. consider a maximum number of $n+1$ electrons per bubble. However, as the $n+1$ bubble phase appears around $\tilde{\nu} = 1/2$, it is unstable against the formation of a stripe phase [81].

This is the famous Wigner crystal [88]. It can be considered a special case of the bubble phase. For higher $\tilde{\nu}$, the system undergoes a succession of phases separated by first order phase transitions with an increasing number of electrons per bubble [87].

From an experimental point of view, little is know about the microscopic structure of the bubble phases. The interpretation as bubbles pinned by disorder is corroborated by the observation of a strongly non-linear $I$-$V$ characteristic, which points to the sudden depinning of the bubble crystal [86]. Apart from that, resonances in the microwave absorption have been interpreted as pinning modes of the bubble crystal in the surrounding disorder potential [89, 90]. Using this technique, the coexistence of the Wigner crystal and the multi-electron bubble phase has been shown over a broad filling factor range, which indicates a first order transition between the respective phases [91].

### *2.5.2 The Stripe Phase*

To unveil the existence of the stripe phase, the longitudinal resistance along two orthogonal crystal directions must be compared. This is done in Fig. 2.18. The measurement was performed on a square-shaped structure in order to obtain a symmetric setup. Figure 2.18 demonstrates that the longitudinal resistance depends strongly on the crystal direction probed in experiment. This observation is most evident at half

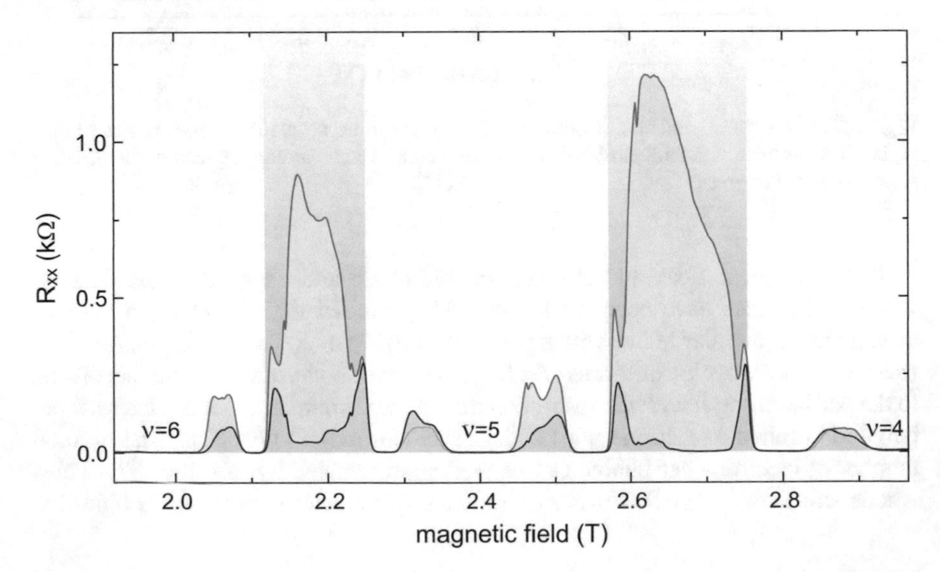


**Fig. 2.18** Longitudinal resistance along two orthogonal current directions (sample temperature ~35 mK). A strong transport anisotropy is visible at each half-filled Landau level branch (marked in *green*), which indicates the formation of a stripe phase

filling of each Landau level branch ($\tilde{\nu} = 1/2$). Since both crystal directions, [110] and [1$\bar{1}$0], are identical in an ideal gallium arsenide crystal, the strong transport anisotropy indicates the existence of an anisotropic, stripe-like electron phase. If the current flows along the stripes ("easy" axis), the resulting voltage drop will be smaller than in the orthogonal direction ("hard" axis). This anisotropy is amplified by an effect called current channeling: If the current is sent perpendicular to the stripes, it gets spread out stronger towards the sample edges causing a higher voltage drop as compared to the orthogonal current direction [92, 93].

The physical mechanism leading to a preferred stripe orientation along specific crystal axes remains an open question up to date. Cooper et al. excluded substrate morphology, steps induced by miscuts as well as an anisotropic effective mass as possible explanations [94]. Instead, piezoelectric effects [95] and a combination of Rashba and Dresselhaus spin-orbit interactions have been proposed [96]. Intriguing is the fact that the stripe orientation depends on the electron density. Below a density of $2.9 \times 10^{11}\,\mathrm{cm}^{-2}$ the easy axis is directed along the [110] direction, whereas for higher densities the [1$\bar{1}$0] direction is preferred [97]. In addition, the stripes can be oriented by applying a magnetic field component in the plane of the 2DES [97–99]. Yet, the exact behavior of the stripes in an in-plane magnetic field seems to depend on the details of the sample [97].

Different theoretical models have been proposed for the microscopic structure of the stripe phase [80]. The charge density wave picture has been mentioned before (Fig. 2.16). In this model, the local filling factor is assumed to be modulated in parallel stripes with a strictly one-dimensional periodicity. Fradkin and Kivelson studied the effects of quantum and thermal fluctuations on the stripe formation and predicted in analogy to liquid crystal behavior the existence of a different class of stripe phases—the electron liquid crystal phases [100]. They are categorized according to their symmetry as smectic, nematic and stripe crystal phase. These phases will be discussed in detail in Chap. 5. Which of the three electron liquid crystal phases is lowest in energy depends on the strength of the shape fluctuations and the average filling factor [100]. As a consequence, different types of stripe phases might be present within the anisotropic region in Fig. 2.18, and transitions between the respective phases may occur as a function of the magnetic field.

Similar to the bubble phase, experimental results providing a detailed microscopic understanding of the stripe phase are scarce. The temperature dependence of the longitudinal resistance points towards the existence of a nematic phase [101, 102]. Recently, the collective modes of the stripe phase at $\nu = 9/2$ were probed in a sophisticated experiment involving surface acoustic waves, microwaves and photoluminescence [103]. In the same experiment, the period of the stripe pattern was determined to $3.6 \cdot R_c$. In addition, pinning mode resonances have been observed also for the stripe phase as reported earlier for the bubble phase [104, 105]. Interestingly, in the case of the stripe phase, a resonance only occurs if the electric field of the microwave signal is oriented along the hard axis.

## 2.6 The Second Landau Level

In the previous sections, it has been shown that the physics in partially filled Landau levels is governed by two classes of competing, interaction-driven phases. At high magnetic fields, when the first Landau level is partially occupied, FQHS are prevalent. In contrast, starting from the third Landau level towards higher filling factors, density-modulated phases, i.e. the bubble and stripe phases, take over. Their existence is heralded either by a reappearance of the IQHE or a strong transport anisotropy. Their dominance over the FQHS in higher Landau levels is rooted in the different exchange and direct Coulomb interactions, which originate from the altered shape and extent of the wavefunction. The competition between density-modulated phases and FQHE is most prominent in the second Landau level. Here, density-modulated phases as well as FQHS coexist, and minute changes of the magnetic field or electron density can lead to transitions between them. This is shown in Fig. 2.19. It displays the longitudinal and Hall resistance between filling factors $\nu = 2$ and $\nu = 4$, measured on a high-quality sample. A large number of different FQHS as well as reentrant states are visible, which emphasizes the superior quality of the sample. The variety of competing phases in the second Landau level is a beautiful example of the enormous richness and complexity of quantum Hall physics. When taking a closer look at Fig. 2.19, it becomes apparent that the second Landau level is exceptional in many regards. Not only the coexistence of density-modulated phases and FQHS is conspicuous but also the appearance of four reentrant states in each branch of the second Landau level leaps to the eye. Moreover, the observation of a fractional quantum Hall state also at half filling of each spin branch is striking ($\nu = {}^5\!/_2$ and $\nu = {}^7\!/_2$). Both points are addressed in more detail hereinafter.

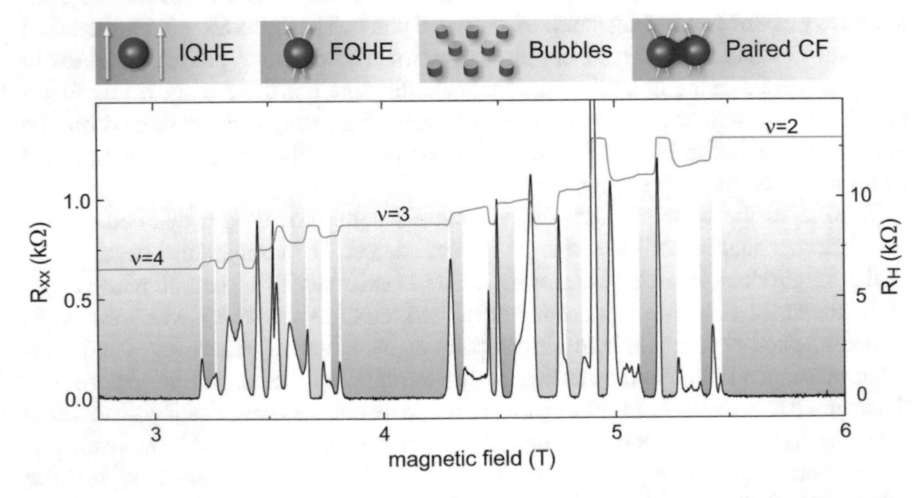


**Fig. 2.19** Longitudinal and Hall resistance in the second Landau level measured on a high-quality sample. Multiple FQHS (marked in *pink*) and reentrant states (marked in *green*) are visible. Remarkable is the appearance of a quantum Hall state at half filling of each spin-split Landau level ($\nu = {}^5\!/_2$ and ${}^7\!/_2$, marked in *blue*)

### 2.6.1 Reentrant States in the Second Landau Level

As mentioned earlier, the second Landau level differs from higher levels in the number of reentrant states. Figure 2.19 shows four reentrant states per spin branch, separated by different FQHS. They appear at partial filling $\tilde{\nu} = 0.29, 0.43, 0.57$ and 0.70 in agreement with previous publications [106–108]. That raises the question whether the reentrant states in the second Landau level are of equal nature as the bubble phases in higher levels. Initially, theories did not consider the second Landau level for bubble phase formation at all [82, 83]. This deficiency was remedied more recently, when the bubble phase consisting of one-electron and two-electron bubbles was shown to be lowest in energy for certain filling factors in the second Landau level [81, 109]. Besides the IQHE transport behavior, other similarities to the bubble phases in higher Landau levels have been observed, such as sharp and hysteretic features in the $I$–$V$ characteristics [106] as well as pinning modes in microwave absorption [110]. In addition, it has been demonstrated that the onset temperatures of the RIQHS in the second Landau level scale with the Coulomb energy, which highlights the importance of electron-electron interactions for the formation of these states [108]. However, also discrepancies to the bubble phases in higher Landau levels are evident. The onset temperatures normalized by the Coulomb energy differ strongly between the second and third Landau level [111]. This observation is in conflict with existing theories and emphasizes that the bubble phases in the second Landau level might be of a more complex nature. Further work is needed to clarify the microscopic structure of the bubble phases in the second Landau level and beyond.

It is important to mention that also the close relative of the bubble phase, the stripe phase, can be observed in the second Landau level. In Fig. 2.19 its position at half filling is taken in by the $\nu = 5/2$ and $7/2$ FQHS. However, tilting the sample with respect to the external magnetic field causes the stripe phase to prevail energetically over the FQHE [98, 99, 112]. The microscopic nature of this tilt-induced stripe phase is the focus of Chap. 5.

### 2.6.2 The $\nu = 5/2$ and $7/2$ States

The observation of FQHS at half fillings comes as a surprise in view of the CF theory introduced in Sect. 2.4.2. There, the partial filling factor $\tilde{\nu} = 1/2$ was described as a compressible Fermi sea of composite fermions. It was concluded that the CF theory only accounts for FQHS at odd-denominator filling factors. Since its discovery in 1987 [113], the $5/2$ state and its particle-hole conjugate, the $7/2$ state, attracted a lot of attention.[5] This is partly due to the fact that these even-denominator FQHS do not comply with the standard CF picture but also because of some interesting theoretical

[5] In the following, we will focus mainly on the $5/2$ state because of its higher stability and therefore greater experimental relevance. Since the $7/2$ state is considered the particle-hole conjugate of the $5/2$ state, all statements should apply equally to the $7/2$ state.

proposals which came up to explain the occurrence of the $^5/_2$ state. In particular the theory of G. Moore and N. Read (together with Greiter et al. [114]) raised a lot of excitement as it ascribes the origin of the $^5/_2$ state to a pairing of two CFs in analogy to the p-wave pairing in a superconductor [115]. By the nature of the p-wave pairing, the resulting state would be spin polarized, and the quasiparticles would have a fractional charge of $e/4$. Even more intriguing is the prediction that the excitations of the Moore-Read state constitute a novel type of quasiparticle called non-abelian anyon. Their exceptional properties are described below. A recent review on the $^5/_2$ state and its potentially non-abelian nature is given in reference [116].

### Non-abelian Anyons and Their Relevance for Topological Quantum Computation

Quasiparticles in the FQHE not only carry fractional charges as mentioned in Sect. 2.4 but also exhibit another fascinating property—they are anyons. Anyons can be considered a generalization of bosons and fermions. They can be distinguished by their exchange statistics. The interchange of anyons results in a general phase factor $e^{i\theta}$ to the wavefunction and is not limited to $-1$ and 1 as in the case of fermions and bosons, respectively [117]. This generalized exchange statistics can only occur in two-dimensional systems. To illustrate this statement, we consider the twofold exchange of two particles with each other. This process is equivalent to describing a closed loop of one particle around the other. When moving a particle around another in a three-dimensional space, the trace can always be deformed continuously down to a point (Fig. 2.20). Therefore, a closed loop of particles is topologically equivalent to no movement at all and multiplies the wavefunction with a phase factor of 1. In this case, the only possible solutions for a single interchange are the factors $-1$ and 1, corresponding to the exchange of fermions and bosons, respectively. In two dimensions, however, a closed loop can no longer be contracted to a point without cutting through the other particle, i.e. the system does not necessarily come back to the same state [118]. Hence, also the phase acquired during particle exchange can have any value and is not bound to 0 and $\pi$ as in the case of fermions and bosons. For the FQHE at $\nu = {}^1/_3$, for instance, a counterclockwise exchange of quasiparticles leads to a phase change of $\pi/3$ [119].

*Non-abelian* anyons are special in the sense that the particle exchange is non-commutative. Suppose we have three particles and want to mutually change their positions (Fig. 2.21). The non-abelian nature brings the system into a different state

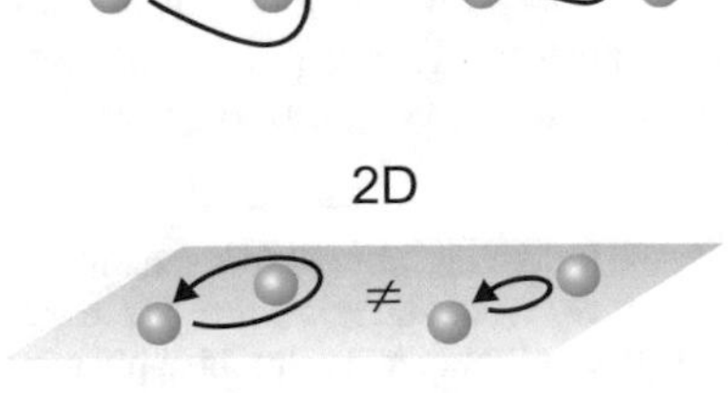

**Fig. 2.20** Particle exchange in two and three dimensions (based on [3])

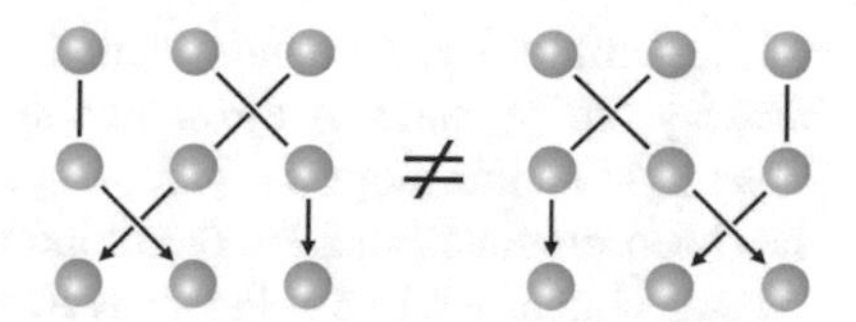

**Fig. 2.21** The exchange of non-abelian particles is not commutative (based on [3])

if first particles on position 1 and 2 interchange and then on positions 2 and 3 in comparison to the particle exchange in the opposite order. Put differently, the operators for particle exchange do not commute. A necessary requirement for non-abelian states of matter is the existence of a highly degenerate ground state which is protected by an energy gap. When particles encircle one another or, in other words, particles are made to "braid", the system transforms from one ground state to another [120]. The transformation is determined solely by the topology of the braiding and is in particular independent of the precise trace taken by the particles. This robustness to local perturbations in combination with the non-abelian braiding statistics provides the basis for topological quantum computation [118–121]. In this particular type of quantum computer, the quantum operations are performed by braiding of non-abelian particles. A topological quantum computer is superior in the sense that it is rather immune to local perturbations of the environment [118]. Standard schemes of quantum computation suffer from the fact that the quantum bits couple to the environment, which causes decoherence and loss of quantum information. In a topological quantum computer local perturbations would affect the path of the particle, but as long as it maintains a loop around the other particle, i.e. as long as the topology is preserved, the quantum information is preserved as well. Instead, the lifetime of quantum information is mainly determined by the sample temperature with respect to the energy gap which protects the degenerate ground state [118, 119].

Besides the mentioned odd-denominator FQHS, other phases of matter are predicted to host non-abelian particles, such as cold atoms [122], p-wave superconductors [118] as well as hybrid systems made out of superconductors in combination with topological insulators [123] or semiconductors [124, 125]. Among these, the most prominently studied candidate for fault-tolerant quantum computation is probably the $^5/_2$ state. In the following, we briefly summarize what is known about the potentially non-abelian nature of the $^5/_2$ state.

### Is the $^5/_2$ State Indeed Home to Non-abelian Anyons?

This question would be answered most convincingly by braiding of $\nu = {}^5/_2$ quasiparticles and thereby testing the underlying exchange statistics. Several proposals for interference experiments at the $^5/_2$ state have been made [126–130], and also first steps towards their realization were undertaken [131, 132]. Even weak signatures of non-abelian anyons were already reported in interferometer structures at $\nu = {}^5/_2$ [133, 134]. The results are, however, difficult to interpret and discussed rather controversially. Further experiments will be necessary to directly prove the non-abelian nature of the $^5/_2$ quasiparticles. Unfortunately, efforts in this direction are hampered by the high fragility of the $^5/_2$ state. The fabrication of gates on top of the sample, a basic requirement to control the interferometer structure, can easily destroy the $5/2$ state. A different way to probe the potentially non-abelian properties might be thermopower [135] or entropy measurements [136]. More precisely, it has been predicted that the thermopower at $\nu = 5/2$ is temperature independent and changes linearly when going away from the exact filling factor if non-abelian anyons are present. Recently, first experiments in this direction have been performed [137]. The results were tentatively claimed to be consistent with the predicted behavior. Even though being still in its infancy, this technique might well prove fruitful in the exploration of the $5/2$ state.

In view of the hurdles which need to be taken to directly measure the non-abelian properties, research in the past has focused stronger on the other predictions of the Moore-Read theory. In particular, the verification of the $e/4$ fractional charge and the full spin polarization would provide strong support for the wavefunction proposed by Moore and Read (sometimes called Pfaffian wavefunction).

The question for the charge of the $5/2$ quasiparticles has been addressed in a number of experiments by different means: tunneling through a quantum point contact [132], noise measurements [138], Aharanov-Bohm interference [133] and experiments using a single-electron transistor [139]. The obtained results are consistently in agreement with a fractional charge of $e/4$. However, establishing the charge $e/4$ does not allow to discriminate between the Pfaffian state and other proposed wavefunctions such as the anti-Pfaffian and the so-called 331 state. The anti-Pfaffian state is similar to the Pfaffian and is also expected to give rise to non-abelian excitations but in addition is particle-hole symmetric, a property lacked by the Pfaffian state [140, 141]. The 331 state was originally proposed by Halperin as a description of a two-component FQHS [51]. It is unpolarized and obeys abelian statistics [142].

Information on the spin polarization of the $5/2$ state would shed further light on its true nature. In view of the verified $e/4$ fractional charge, a polarized state would indirectly provide strong support for the non-abelian character of the $5/2$ quasiparticles. First experiments on the spin polarization examined the behavior of the $5/2$ state when tilting the sample with respect to the magnetic field [143]. This selectively enhances the Zeeman energy relative to the Coulomb energy since the latter only depends on the perpendicular magnetic field component. The disappearance of the $5/2$ state under tilt was therefore taken as an indication for an unpolarized state. In later years, the vanishing of the $5/2$ state was attributed rather to the emergence of an anisotropic stripe phase under the influence of the in-plane magnetic field [98, 99, 112, 144, 145]. Pan et al. approached the spin polarization of the $5/2$ state in a different way [146]. They determined the energy gap at $\nu = 5/2$ over a wide density range and compared its evolution with the behavior of the FQHS at $\nu = 8/5$, a state known to be unpolarized at low electron densities. From the observation that the $8/5$ state undergoes a spin transition whereas the $5/2$ state remains mostly unaffected, the authors concluded a spin-polarized state at $\nu = 5/2$. In contrast, a more recent study of the $\nu = 5/2$ energy gap at various perpendicular and in-plane magnetic fields came to the conclusion that their data is more consistent with a spin-unpolarized state [147]. Besides transport and tilted-field experiments, also other techniques have been used to assess the spin polarization at $\nu = 5/2$. Rhone et al. probed the $5/2$ state by inelastic light scattering and ruled out a fully polarized state from the absence of a long-wavelength spin wave mode [148]. However, shortly after, such a spin mode was found by Wurstbauer et al. [149]. Apart from that, Stern et al. claimed the observation of an unpolarized $^5/_2$ state using photoluminescence spectroscopy [150]. Yet, the difficulty behind optical experiments is that the excitation of electron-hole pairs might disturb the electron system and may impair the proper assignment of the filling factor. A different access to the spin polarization provides nuclear magnetic resonance spectroscopy. It uses a characteristic shift of the nuclear resonance frequency to detect the presence of a spin-polarized electron system. This technique has been employed in two similar studies, both of which found a fully spin-polarized state [151, 152].

After all, contradictory conclusions from different experiments keep the issue of the spin polarization at $\nu = {}^5/_2$ open [116]. Since the discrepancies between the different experiments might arise from the fragility of the $^5/_2$ state, it is required to repeat the measurement of the spin polarization on samples with a higher quality showing a well-developed $^5/_2$ state. This task will be addressed in Chap. 4.

## References


1. J.H. Davies, *The Physics of Low-Dimensional Semiconductors* (Cambridge University Press, Cambridge, 1998)
2. J.K. Jain, *Composite Fermions* (Cambridge University Press, Cambridge, 2007)
3. J. Nübler, Density dependence of the $\nu = 5/2$ fractional quantum Hall effect - Compressibility of a two-dimensional electron system under microwave irradiation. Ph.D. thesis (Eberhard Karls Universität, Tübingen, 2011)
4. J. Göres, Correlation effects in 2-dimensional electron systems - Composite fermions and electron liquid crystals. Ph.D. thesis (Universität Stuttgart, 2004)
5. X. Huang, Critical phenomena in bilayer excitonic condensates. Ph.D. thesis (Universität Stuttgart, 2012)
6. O. Stern, Spin phenomena in the fractional quantum Hall effect: NMR and magnetotransport studies. Ph.D. thesis (Universität Stuttgart, 2005)
7. N. Freytag, The electron spin polarization in the lowest Landau level. Ph.D. thesis (Université Joseph Fourier, Grenoble, 2001)
8. C. Dean, A study of the fractional quantum Hall energy gap at half filling. Ph.D. thesis (McGill University, Montréal, 2008)
9. S. Birner, T. Zibold, T. Andlauer, T. Kubis, M. Sabathil, A. Trellakis, P. Vogl, Nextnano: General purpose 3-D simulations. IEEE Trans. Electron Devices **54**, 2137 (2007)
10. V. Umansky, M. Heiblum, Y. Levinson, J. Smet, J. Nübler, M. Dolev, MBE growth of ultra-low disorder 2DEG with mobility exceeding $35 \times 10^6$ cm$^2$/Vs. J. Cryst. Growth **311**, 1658 (2009)
11. R. Dingle, H.L. Störmer, A.C. Gossard, W. Wiegmann, Electron mobilities in modulation-doped semiconductor heterojunction superlattices. Appl. Phys. Lett. **33**, 665 (1978)
12. E.H. Hwang, S. Das Sarma, Limit to two-dimensional mobility in modulation-doped GaAs quantum structures: How to achieve a mobility of 100 million. Phys. Rev. B **77**, 235437 (2008)
13. G. Gamez, K. Muraki, $\nu = 5/2$ fractional quantum Hall state in low-mobility electron systems: Different roles of disorder. Phys. Rev. B **88**, 075308 (2013)
14. K.-J. Friedland, R. Hey, H. Kostial, R. Klann, K. Ploog, New concept for the reduction of impurity scattering in remotely doped GaAs quantum wells. Phys. Rev. Lett. **77**, 4616 (1996)

15. T. Baba, T. Mizutani, M. Ogawa, Elimination of persistent photoconductivity and improvement in Si activation coefficient by Al spatial separation from Ga and Si in Al-Ga-As: Si solid system - a novel short period AlAs/n-GaAs superlattice -. Jpn. J. Appl. Phys. **22**, 627 (1983)
16. P.M. Mooney, Deep donor levels (DX centers) in III-V semiconductors. J. Appl. Phys. **67**, R1 (1990)
17. P. Drude, Zur Elektronentheorie der Metalle (I.). Ann. Phys. **306**, 566 (1900)
18. P. Drude, Zur Elektronentheorie der Metalle (II.). Ann. Phys. **308**, 369 (1900)
19. E.H. Hall, On a new action of the magnet on electric currents. Am. J. Math. **2**, 287 (1879)
20. G.F. Giuliani, G. Vignale, *Quantum Theory of the Electron Liquid* (Cambridge University Press, Cambridge, 2005)
21. Z.F. Ezawa, *Quantum Hall Effects* (World Scientific Publishing Co. Pte. Ltd., Singapore, 2008)
22. A.M. White, I. Hinchliffe, P.J. Dean, P.D. Greene, Zeeman spectra of the principal bound exciton in Sn-doped gallium arsenide. Solid State Commun. **10**, 497 (1972)
23. C. Weisbuch, C. Hermann, Optical detection of conduction-electron spin resonance in GaAs, $Ga_{1-x}In_xAs$ and $Ga_{1-x}Al_xAs$. Phys. Rev. B **15**, 816 (1977)
24. T.P. Smith, B.B. Goldberg, P.J. Stiles, M. Heiblum, Direct measurement of the density of states of a two-dimensional electron gas. Phys. Rev. Lett. **32**, 2696 (1985)
25. J.P. Eisenstein, H.L. Stormer, V. Narayanamurti, A.Y. Cho, A.C. Gossard, C.W. Tu, Density of states and de Haas-van Alphen effect in two-dimensional electron systems. Phys. Rev. Lett. **55**, 875 (1985)
26. V. Mosser, D. Weiss, K. von Klitzing, K. Ploog, G. Weimann, Density of states of GaAs-AlGaAs-heterostructures deduced from temperature dependent magnetocapacitance measurements. Solid State Commun. **58**, 5 (1986)
27. R.C. Ashoori, R.H. Silsbee, The Landau level density of states as a function of Fermi energy in the two dimensional electron gas. Solid State Commun. **81**, 821 (1992)
28. A. Potts, R. Shepherd, W.G. Herrenden-Harker, M. Elliott, C.L. Jones, A. Usher, G.A.C. Jones, D.A. Ritchie, E.H. Linfield, M. Grimshaw, Magnetization studies of Landau level broadening in two-dimensional electron systems. J. Phys. Condens. Matter **8**, 5189 (1996)
29. S. Ilani, J. Martin, E. Teitelbaum, J.H. Smet, D. Mahalu, V. Umansky, A. Yacoby, The microscopic nature of localization in the quantum Hall effect. Nature **427**, 328 (2004)
30. M. Büttiker, Absence of backscattering in the quantum Hall effect in multiprobe conductors. Phys. Rev. B **38**, 9375 (1988)
31. R. Landauer, Electrical transport in open and closed systems. Z. Phys. B **68**, 217 (1987)
32. D.B. Chklovskii, B.I. Shklovskii, L.I. Glazman, Electrostatics of edge channels. Phys. Rev. B **46**, 4026 (1992)
33. D.B. Chklovskii, K.A. Matveev, B.I. Shklovskii, Ballistic conductance of interacting electrons in the quantum Hall regime. Phys. Rev. B **47**, 605 (1993)
34. A. Siddiki, R.R. Gerhardts, Incompressible strips in dissipative Hall bars as origin of quantized Hall plateaus. Phys. Rev. B **70**, 195335 (2004)
35. K. von Klitzing, R. Gerhardts, J. Weis, 25 Jahre Quanten-Hall-Effekt. Phys. J. **4**, 37 (2005)
36. E. Ahlswede, P. Weitz, J. Weis, K. von Klitzing, K. Eberl, Hall potential profiles in the quantum Hall regime measured by a scanning force microscope. Phys. B **298**, 562 (2001)
37. E. Ahlswede, J. Weis, K. von Klitzing, K. Eberl, Hall potential distribution in the quantum Hall regime in the vicinity of a potential probe contact. Phys. E **12**, 165 (2002)
38. J. Weis, Y.Y. Wei, K. von Klitzing, Single-electron transistor probes two-dimensional electron system in high magnetic fields. Phys. E **3**, 23 (1998)
39. Y.Y. Wei, J. Weis, K. von Klitzing, K. Eberl, Edge strips in the quantum Hall regime imaged by a single-electron transistor. Phys. Rev. Lett. **81**, 1674 (1998)
40. E. Ahlswede, Potential- und Stromverteilung beim Quanten-Hall-Effekt bestimmt mittels Rasterkraftmikroskopie. Ph.D. thesis (Universität Stuttgart, 2002)
41. D.C. Tsui, H.L. Stormer, A.C. Gossard, Two-dimensional magnetotransport in the extreme quantum limit. Phys. Rev. Lett. **48**, 1559 (1982)

42. K. von Klitzing, G. Dorda, M. Pepper, New method for high-accuracy determination of the fine-structure constant based on quantized Hall resistance. Phys. Rev. Lett. **45**, 494 (1980)
43. W. Pan, J.S. Xia, H.L. Stormer, D.C. Tsui, C. Vicente, E.D. Adams, N.S. Sullivan, L.N. Pfeiffer, K.W. Baldwin, K.W. West, Experimental studies of the fractional quantum Hall effect in the first excited Landau level. Phys. Rev. B **77**, 075307 (2008)
44. R.B. Laughlin, Anomalous quantum Hall effect: An incompressible quantum fluid with fractionally charged excitations. Phys. Rev. Lett. **50**, 1395 (1983)
45. F.D.M. Haldane, Fractional quantization of the Hall effect: A hierarchy of incompressible quantum fluid states. Phys. Rev. Lett. **51**, 605 (1983)
46. B.I. Halperin, Statistics of quasiparticles and the hierarchy of fractional quantized Hall states. Phys. Rev. Lett. **52**, 1583 (1984)
47. S.M. Girvin, A.H. MacDonald, Off-diagonal long-range order, oblique confinement, and the fractional quantum Hall effect. Phys. Rev. Lett. **58**, 1252 (1987)
48. S.C. Zhang, T.H. Hansson, S. Kivelson, Effective-field-theory model for the fractional quantum Hall effect. Phys. Rev. Lett. **62**, 82 (1989)
49. N. Read, Order parameter and Ginzburg-Landau theory for the fractional quantum Hall effect. Phys. Rev. Lett. **62**, 86 (1989)
50. R.B. Laughlin, Nobel lecture: Fractional quantization. Rev. Mod. Phys. **71**, 863 (1999)
51. B.I. Halperin, Theory of the quantized Hall conductance. Helv. Phys. Acta **56**, 75 (1983)
52. H.L. Störmer, The fractional quantum Hall effect, in *Nobel Lectures in Physics (1996-2000)* (World Scientific Publishing Co. Pte. Ltd., Singapore, 2002), pp. 295–325
53. V.J. Goldman, B. Su, Resonant tunneling in the quantum Hall regime: Measurement of fractional charge. Science **267**, 1010 (1995)
54. L. Saminadayar, D.C. Glattli, Y. Jin, B. Etienne, Observation of the e/3 fractionally charged Laughlin quasiparticle. Phys. Rev. Lett. **79**, 2526 (1997)
55. R. de Picciotto, M. Reznikov, M. Heiblum, V. Umansky, G. Bunin, D. Mahalu, Direct observation of a fractional charge. Nature **389**, 162 (1997)
56. M. Reznikov, R. de Picciotto, T.G. Griffiths, M. Heiblum, V. Umansky, Observation of quasiparticles with one-fifth of an electron's charge. Nature **399**, 238 (1999)
57. J. Martin, S. Ilani, B. Verdene, J. Smet, V. Umansky, D. Mahalu, D. Schuh, G. Abstreiter, A. Yacoby, Localization of fractionally charged quasi-particles. Science **305**, 980 (2004)
58. S.T. Bramwell, S.R. Giblin, S. Calder, R. Aldus, D. Prabhakaran, T. Fennell, Measurement of the charge and current of magnetic monopoles in spin ice. Nature **461**, 956 (2009)
59. T. Fennell, P.P. Deen, A.R. Wildes, K. Schmalzl, D. Prabhakaran, A.T. Boothroyd, R.J. Aldus, D.F. McMorrow, S.T. Bramwell, Magnetic coulomb phase in the spin ice $Ho_2Ti_2O_7$. Science **326**, 415 (2009)
60. D.J.P. Morris, D.A. Tennant, S.A. Grigera, B. Klemke, C. Castelnovo, R. Moessner, C. Czternasty, M. Meissner, K.C. Rule, J.-U. Hoffmann, K. Kiefer, S. Gerischer, D. Slobinsky, R.S. Perry, Dirac strings and magnetic monopoles in the spin ice $Dy_2Ti_2O_7$. Science **326**, 411 (2009)
61. J. Schlappa, K. Wohlfeld, K.J. Zhou, M. Mourigal, M.W. Haverkort, V.N. Strocov, L. Hozoi, C. Monney, S. Nishimoto, S. Singh, A. Revcolevschi, J.-S. Caux, L. Patthey, H.M. Rønnow, J. van den Brink, T. Schmitt, Spin-orbital separation in the quasi-one-dimensional Mott insulator $Sr_2CuO_3$. Nature **485**, 82 (2012)
62. C. Kim, A.Y. Matsuura, Z.-X. Shen, N. Motoyama, H. Eisaki, S. Uchida, T. Tohyama, S. Maekawa, Observation of spin-charge separation in one-dimensional $SrCuO_2$. Phys. Rev. Lett. **77**, 4054 (1996)
63. Y. Jompol, C.J.B. Ford, J.P. Griffiths, I. Farrer, G.A.C. Jones, D. Anderson, D.A. Ritchie, T.W. Silk, A.J. Schofield, Probing spin-charge separation in a Tomonaga-Luttinger liquid. Science **325**, 597 (2009)
64. F.D.M. Haldane, E.H. Rezayi, Finite-size studies of the incompressible state of the fractionally quantized Hall effect and its excitations. Phys. Rev. Lett. **54**, 237 (1985)
65. R. Morf, B.I. Halperin, Monte Carlo evaluation of trial wave functions for the fractional quantized Hall effect: Disk geometry. Phys. Rev. B **33**, 2221 (1986)

66. S.M. Girvin, Particle-hole symmetry in the anomalous quantum Hall effect. Phys. Rev. B **29**, 6012 (1984)
67. J.K. Jain, Composite-fermion approach for the fractional quantum Hall effect. Phys. Rev. Lett. **63**, 199 (1989)
68. R.L. Willett, M.A. Paalanen, R.R. Ruel, K.W. West, L.N. Pfeiffer, D.J. Bishop, Anomalous sound propagation at $\nu = 1/2$ in a 2D electron gas: Observation of a spontaneously broken translational symmetry? Phys. Rev. Lett. **65**, 112 (1990)
69. R.R. Du, H.L. Stormer, D.C. Tsui, L.N. Pfeiffer, K.W. West, Experimental evidence for new particles in the fractional quantum Hall effect. Phys. Rev. Lett. **70**, 2944 (1993)
70. R.R. Du, A.S. Yeh, H.L. Stormer, D.C. Tsui, L.N. Pfeiffer, K.W. West, Fractional quantum Hall effect around $\nu = 3/2$: Composite fermions with a spin. Phys. Rev. Lett. **75**, 3926 (1995)
71. R.L. Willett, R.R. Ruel, M.A. Paalanen, K.W. West, L.N. Pfeiffer, Enhanced finite-wave-vector conductivity at multiple even-denominator filling factors in two-dimensional electron systems. Phys. Rev. B **47**, 7344 (1993)
72. R.L. Willett, R.R. Ruel, K.W. West, L.N. Pfeiffer, Experimental demonstration of a Fermi surface at one-half filling of the lowest Landau level. Phys. Rev. Lett. **71**, 3846 (1993)
73. W. Kang, H.L. Stormer, L.N. Pfeiffer, K.W. Baldwin, K.W. West, How real are composite fermions? Phys. Rev. Lett. **71**, 3850 (1993)
74. V.J. Goldman, B. Su, J.K. Jain, Detection of composite fermions by magnetic focusing. Phys. Rev. Lett. **72**, 2065 (1994)
75. J.H. Smet, D. Weiss, R.H. Blick, G. Lütjering, K. von Klitzing, R. Fleischmann, R. Ketzmerick, T. Geisel, G. Weimann, Magnetic focusing of composite fermions through arrays of cavities. Phys. Rev. Lett. **77**, 2272 (1996)
76. I.V. Kukushkin, J.H. Smet, K. von Klitzing, W. Wegscheider, Cyclotron resonance of composite fermions. Nature **415**, 409 (2002)
77. W. Pan, H.L. Stormer, D.C. Tsui, L.N. Pfeiffer, K.W. Baldwin, K.W. West, Fractional quantum Hall effect of composite fermions. Phys. Rev. Lett. **90**, 016801 (2003)
78. J.K. Jain, The composite fermion: A quantum particle and its quantum fluids. Phys. Today **53**, 39 (2000)
79. M.M. Fogler, A.A. Koulakov, Laughlin liquid to charge-density-wave transition at high Landau levels. Phys. Rev. B **55**, 9326 (1997)
80. M.M. Fogler, Stripe and bubble phases in quantum Hall systems, in *High Magnetic Fields: Applications in Condensed Matter Physics and Spectroscopy* (Springer, Berlin, 2002), pp. 98–138
81. M.O. Goerbig, P. Lederer, C. Morais Smith, Competition between quantum-liquid and electron-solid phases in intermediate Landau levels. Phys. Rev. B **69**, 115327 (2004)
82. M.M. Fogler, A.A. Koulakov, B.I. Shklovskii, Ground state of a two-dimensional electron liquid in a weak magnetic field. Phys. Rev. B **54**, 1853 (1996)
83. A.A. Koulakov, M.M. Fogler, B.I. Shklovskii, Charge density wave in two-dimensional electron liquid in weak magnetic field. Phys. Rev. Lett. **76**, 499 (1996)
84. R.R. Du, D.C. Tsui, H.L. Stormer, L.N. Pfeiffer, K.W. Baldwin, K.W. West, Strongly anisotropic transport in higher two-dimensional Landau levels. Solid State Commun. **109**, 389 (1999)
85. M.P. Lilly, K.B. Cooper, J.P. Eisenstein, L.N. Pfeiffer, K.W. West, Evidence for an anisotropic state of two-dimensional electrons in high Landau levels. Phys. Rev. Lett. **82**, 394 (1999)
86. K.B. Cooper, M.P. Lilly, J.P. Eisenstein, L.N. Pfeiffer, K.W. West, Insulating phases of two-dimensional electrons in high Landau levels: Observation of sharp thresholds to conduction. Phys. Rev. B **60**, 285 (1999)
87. R. Côté, C.B. Doiron, J. Bourassa, H.A. Fertig, Dynamics of electrons in quantum Hall bubble phases. Phys. Rev. B **68**, 155327 (2003)
88. E. Wigner, On the interaction of electrons in metals. Phys. Rev. **46**, 1002 (1934)
89. R. Lewis, P.D. Ye, L.W. Engel, D.C. Tsui, L.N. Pfeiffer, K.W. West, Microwave resonance of the bubble phases in 1/4 and 3/4 filled high Landau levels. Phys. Rev. Lett. **89**, 136804 (2002)

90. P.D. Ye, L.W. Engel, D.C. Tsui, R.M. Lewis, L.N. Pfeiffer, K.W. West, Correlation lengths of the Wigner-crystal order in a two-dimensional electron system at high magnetic fields. Phys. Rev. Lett. **89**, 176802 (2002)
91. R.M. Lewis, Y. Chen, L.W. Engel, D.C. Tsui, P.D. Ye, L.N. Pfeiffer, K.W. West, Evidence of a first-order phase transition between Wigner-crystal and bubble phases of 2D electrons in higher Landau levels. Phys. Rev. Lett. **93**, 176808 (2004)
92. S.H. Simon, Comment on "Evidence for an anisotropic state of two-dimensional electrons in high Landau levels". Phys. Rev. Lett. **83**, 4223 (1999)
93. R.L. Willett, K.W. West, L.N. Pfeiffer, Current-path properties of the transport anisotropy at filling factor 9/2. Phys. Rev. Lett. **87**, 196805 (2001)
94. K.B. Cooper, M.P. Lilly, J.P. Eisenstein, T. Jungwirth, L.N. Pfeiffer, K.W. West, An investigation of orientational symmetry-breaking mechanisms in high Landau levels. Solid State Commun. **119**, 89 (2001)
95. D.V. Fil, Piezoelectric mechanism for the orientation of stripe structures in two-dimensional electron systems. Low Temp. Phys. **55**, 1 (2000)
96. I. Sodemann, A.H. Macdonald, Theory of orientational pinning in quantum Hall nematics. arXiv:1307.5489 (2013)
97. J. Zhu, W. Pan, H.L. Stormer, L.N. Pfeiffer, K.W. West, Density-induced interchange of anisotropy axes at half-filled high Landau levels. Phys. Rev. Lett. **88**, 116803 (2002)
98. M.P. Lilly, K.B. Cooper, J.P. Eisenstein, L.N. Pfeiffer, K.W. West, Anisotropic states of two-dimensional electron systems in high Landau levels: Effect of an in-plane magnetic field. Phys. Rev. Lett. **83**, 824 (1999)
99. W. Pan, R.R. Du, H.L. Stormer, D.C. Tsui, L.N. Pfeiffer, K.W. Baldwin, K.W. West, Strongly anisotropic electronic transport at Landau level filling factor $\nu = 9/2$ and $\nu = 5/2$ under a tilted magnetic field. Phys. Rev. Lett. **83**, 820 (1999)
100. E. Fradkin, S.A. Kivelson, Liquid-crystal phases of quantum Hall systems. Phys. Rev. B **59**, 8065 (1999)
101. E. Fradkin, S.A. Kivelson, E. Manousakis, K. Nho, Nematic phase of the two-dimensional electron gas in a magnetic field. Phys. Rev. Lett. **84**, 1982 (2000)
102. K.B. Cooper, M.P. Lilly, J.P. Eisenstein, L.N. Pfeiffer, K.W. West, Onset of anisotropic transport of two-dimensional electrons in high Landau levels: Possible isotropic-to-nematic liquid-crystal phase transition. Phys. Rev. B **65**, 241313 (2002)
103. I.V. Kukushkin, V. Umansky, K. von Klitzing, J.H. Smet, Collective modes and the periodicity of quantum Hall stripes. Phys. Rev. Lett. **106**, 206804 (2011)
104. G. Sambandamurthy, R.M. Lewis, H. Zhu, Y.P. Chen, L.W. Engel, D.C. Tsui, L.N. Pfeiffer, K.W. West, Observation of pinning mode of stripe phases of 2D systems in high Landau levels. Phys. Rev. Lett. **100**, 256801 (2008)
105. H. Zhu, G. Sambandamurthy, L.W. Engel, D.C. Tsui, L.N. Pfeiffer, K.W. West, Pinning mode resonances of 2D electron stripe phases: Effect of an in-plane magnetic field. Phys. Rev. Lett. **102**, 136804 (2009)
106. J.P. Eisenstein, K.B. Cooper, L.N. Pfeiffer, K.W. West, Insulating and fractional quantum Hall states in the first excited Landau level. Phys. Rev. Lett. **88**, 076801 (2002)
107. J.S. Xia, W. Pan, C.L. Vicente, E.D. Adams, N.S. Sullivan, H.L. Stormer, D.C. Tsui, L.N. Pfeiffer, K.W. Baldwin, K.W. West, Electron correlation in the second Landau level: A competition between many nearly degenerate quantum phases. Phys. Rev. Lett. **93**, 176809 (2004)
108. N. Deng, A. Kumar, M.J. Manfra, L.N. Pfeiffer, K.W. West, G.A. Csáthy, Collective nature of the reentrant integer quantum Hall states in the second Landau level. Phys. Rev. Lett. **108**, 086803 (2012)
109. M.O. Goerbig, P. Lederer, C. Morais Smith, Microscopic theory of the reentrant integer quantum Hall effect in the first and second excited Landau levels. Phys. Rev. B **68**, 241302 (2003)
110. R.M. Lewis, Y.P. Chen, L.W. Engel, D.C. Tsui, L.N. Pfeiffer, K.W. West, Microwave resonance of the reentrant insulating quantum Hall phases in the first excited Landau level. Phys. Rev. B **71**, 081301 (2005)

111. N. Deng, J.D. Watson, L.P. Rokhinson, M.J. Manfra, G.A. Csáthy, Contrasting energy scales of the reentrant integer quantum Hall states. Phys. Rev. B **86**, 201301 (2012)
112. J. Xia, V. Cvicek, J.P. Eisenstein, L.N. Pfeiffer, K.W. West, Tilt-induced anisotropic to isotropic phase transition at $\nu = 5/2$. Phys. Rev. Lett. **105**, 176807 (2010)
113. R. Willett, J.P. Eisenstein, H.L. Störmer, D.C. Tsui, A.C. Gossard, J.H. English, Observation of an even-denominator quantum number in the fractional quantum Hall effect. Phys. Rev. Lett. **59**, 1776 (1987)
114. M. Greiter, X.G. Wen, F. Wilczek, Paired Hall states. Nucl. Phys. B **374**, 567 (1992)
115. G. Moore, N. Read, Nonabelions in the fractional quantum Hall effect. Nucl. Phys. B **360**, 362 (1991)
116. R.L. Willett, The quantum Hall effect at 5/2 filling factor. Rep. Prog. Phys. **76**, 076501 (2013)
117. F. Wilczek, *Fractional Statistics and Anyon Superconductivity* (World Scientific Publishing Co. Pte. Ltd., Singapore, 1990)
118. C. Nayak, S.H. Simon, A. Stern, M. Freedman, S. Das Sarma, Non-abelian anyons and topological quantum computation. Rev. Mod. Phys. **80**, 1083 (2008)
119. S. Das Sarma, M. Freedman, C. Nayak, Topological Quantum Computation. Phys. Today **59**, 32 (2006)
120. A. Stern, Non-abelian states of matter. Nature **464**, 187 (2010)
121. A.Y. Kitaev, Fault-tolerant quantum computation by anyons. Ann. Phys. **303**, 2 (2003)
122. N.R. Cooper, N.K. Wilkin, J.M.F. Gunn, Quantum phases of vortices in rotating Bose-Einstein condensates. Phys. Rev. Lett. **87**, 120405 (2001)
123. L. Fu, C.L. Kane, Superconducting proximity effect and Majorana fermions at the surface of a topological insulator. Phys. Rev. Lett. **100**, 096407 (2008)
124. J.D. Sau, R.M. Lutchyn, S. Tewari, S. Das Sarma, Generic new platform for topological quantum computation using semiconductor heterostructures. Phys. Rev. Lett. **104**, 40502 (2010)
125. R.M. Lutchyn, J.D. Sau, S. Das Sarma, Majorana fermions and a topological phase transition in semiconductor-superconductor heterostructures. Phys. Rev. Lett. **105**, 077001 (2010)
126. S. Das Sarma, M. Freedman, C. Nayak, Topologically protected qubits from a possible non-abelian fractional quantum Hall state. Phys. Rev. Lett. **94**, 166802 (2005)
127. D.E. Feldman, A. Kitaev, Detecting non-abelian statistics with an electronic Mach-Zehnder interferometer. Phys. Rev. Lett. **97**, 186803 (2006)
128. A. Stern, B.I. Halperin, Proposed experiments to probe the non-abelian $\nu = 5/2$ quantum Hall state. Phys. Rev. Lett. **96**, 016802 (2006)
129. C.-Y. Hou, C. Chamon, Wormhole' geometry for entrapping topologically protected qubits in non-abelian quantum Hall states and probing them with voltage and noise measurements. Phys. Rev. Lett. **97**, 146802 (2006)
130. P. Bonderson, A. Kitaev, K. Shtengel, Detecting non-abelian statistics in the $\nu = 5/2$ fractional quantum Hall state. Phys. Rev. Lett. **96**, 016803 (2006)
131. J.B. Miller, I.P. Radu, D.M. Zumbühl, E.M. Levenson-Falk, M.A. Kastner, C.M. Marcus, L.N. Pfeiffer, K.W. West, Fractional quantum Hall effect in a quantum point contact at filling fraction 5/2. Nat. Phys. **3**, 561 (2007)
132. I.P. Radu, J.B. Miller, C.M. Marcus, M.A. Kastner, L.N. Pfeiffer, K.W. West, Quasi-particle properties from tunneling in the $\nu = 5/2$ fractional quantum Hall state. Science **320**, 899 (2008)
133. R.L. Willett, L.N. Pfeiffer, K.W. West, Alternation and interchange of e/4 and e/2 period interference oscillations consistent with filling factor 5/2 non-abelian quasiparticles. Phys. Rev. B **82**, 205301 (2010)
134. R.L. Willett, C. Nayak, K. Shtengel, L.N. Pfeiffer, K.W. West, Magnetic field-tuned Aharonov-Bohm oscillations and evidence for non-abelian anyons at $\nu = 5/2$. Phys. Rev. Lett. **111**, 186401 (2013)
135. K. Yang, B.I. Halperin, Thermopower as a possible probe of non-abelian quasiparticle statistics in fractional quantum Hall liquids. Phys. Rev. B **79**, 115317 (2009)

136. N.R. Cooper, A. Stern, Observable bulk signatures of non-abelian quantum Hall states. Phys. Rev. Lett. **102**, 176807 (2009)
137. W.E. Chickering, J.P. Eisenstein, L.N. Pfeiffer, K.W. West, Thermoelectric response of fractional quantized Hall and reentrant insulating states in the N=1 Landau level. Phys. Rev. B **87**, 075302 (2013)
138. M. Dolev, M. Heiblum, V. Umansky, A. Stern, D. Mahalu, Observation of a quarter of an electron charge at the $\nu = 5/2$ quantum Hall state. Nature **452**, 829 (2008)
139. V. Venkatachalam, A. Yacoby, L. Pfeiffer, K. West, Local charge of the $\nu = 5/2$ fractional quantum Hall state. Nature **469**, 185 (2011)
140. M. Levin, B.I. Halperin, B. Rosenow, Particle-hole symmetry and the Pfaffian state. Phys. Rev. Lett. **99**, 236806 (2007)
141. S.-S. Lee, S. Ryu, C. Nayak, M.P.A. Fisher, Particle-hole symmetry and the $\nu = 5/2$ quantum Hall state. Phys. Rev. Lett. **99**, 236807 (2007)
142. N. Read, E. Rezayi, Quasiholes and fermionic zero modes of paired fractional quantum Hall states: The mechanism for non-abelian statistics. Phys. Rev. B **54**, 864 (1996)
143. J.P. Eisenstein, R. Willett, H.L. Stormer, D.C. Tsui, A.C. Gossard, J.H. English, Collapse of the even-denominator fractional quantum Hall effect in tilted fields. Phys. Rev. Lett. **61**, 997 (1988)
144. R.H. Morf, Transition from quantum Hall to compressible states in the second Landau level: New light on the $\nu = 5/2$ enigma. Phys. Rev. Lett. **80**, 1505 (1998)
145. E.H. Rezayi, F.D.M. Haldane, Incompressible paired Hall state, stripe order, and the composite fermion liquid phase in half-filled Landau levels. Phys. Rev. Lett. **84**, 4685 (2000)
146. W. Pan, H.L. Stormer, D.C. Tsui, L.N. Pfeiffer, K.W. Baldwin, K.W. West, Experimental evidence for a spin-polarized ground state in the $\nu = 5/2$ fractional quantum Hall effect. Solid State Commun. **119**, 641 (2001)
147. S. Das Sarma, G. Gervais, X. Zhou, Energy gap and spin polarization in the 5/2 fractional quantum Hall effect. Phys. Rev. B **82**, 115330 (2010)
148. T.D. Rhone, J. Yan, Y. Gallais, A. Pinczuk, L. Pfeiffer, K. West, Rapid collapse of spin waves in nonuniform phases of the second Landau level. Phys. Rev. Lett. **106**, 196805 (2011)
149. U. Wurstbauer, K.W. West, L.N. Pfeiffer, A. Pinczuk, Resonant inelastic light scattering investigation of low-lying gapped excitations in the quantum fluid at $\nu = 5/2$. Phys. Rev. Lett. **110**, 026801 (2013)
150. M. Stern, P. Plochocka, V. Umansky, D.K. Maude, M. Potemski, I. Bar-Joseph, Optical probing of the spin polarization of the $\nu = 5/2$ quantum Hall state. Phys. Rev. Lett. **105**, 096801 (2010)
151. M. Stern, B.A. Piot, Y. Vardi, V. Umansky, P. Plochocka, D.K. Maude, I. Bar-Joseph, NMR probing of the spin polarization of the $\nu = 5/2$ quantum Hall state. Phys. Rev. Lett. **108**, 066810 (2012)
152. L. Tiemann, G. Gamez, N. Kumada, K. Muraki, Unraveling the spin polarization of the $\nu = 5/2$ fractional quantum Hall state. Science **335**, 828 (2012)

# Chapter 3
# Electron–Nuclear Spin Interaction in the Quantum Hall Regime

The spin degree of freedom only played a minor role in the introduction to the integer and fractional quantum Hall effect given in the previous chapter. It is not essential to understand the formation of quantum Hall states. However, when taking a deeper look into quantum Hall physics, the importance of spin related phenomena becomes evident. Today, many interesting spin phenomena are known to exist, such as spin transitions of FQHS [1–3], ferromagnetic spin ordering [3–6] and skyrmionic spin excitations [7, 8]. The spin degree of freedom enriches the quantum Hall physics in many regards. Its complexity increases by the coupling of electron spins to the nuclear spin system established via the hyperfine interaction. Even though this effect is rather weak, it significantly alters the electronic spin properties. The interaction between the electronic and nuclear spin system as well as its utilization to investigate electronic spin excitations is the main focus of this chapter. In the last part of this chapter, we expand on a different but closely related topic, namely the filling factor dependence of the nuclear spin polarization.

## 3.1 Introduction to the Spin Physics in the Quantum Hall Regime

Before embarking on the experimental sections, first a brief overview on the electronic and nuclear spin system as well as the hyperfine coupling between these two systems is given. Detailed information can be found in references [9, 10].

### *3.1.1 The Role of the Electron Spin System*

The physics of electron spins in the quantum Hall effect is dominated by the Zeeman splitting of the Landau levels. All electrons within a 2DES are distributed in



B. Frieß, *Spin and Charge Ordering in the Quantum Hall Regime*, Springer Theses, DOI 10.1007/978-3-319-33536-0_3

consecutive energy levels of opposite spin orientation (Fig. 2.9). Therefore, the spin polarization oscillates when increasing the magnetic field. The overall spin polarization $\mathcal{P}$ is generally defined as

$$\mathcal{P} = \frac{n_\uparrow - n_\downarrow}{n_\uparrow + n_\downarrow}, \tag{3.1}$$

where $n_\uparrow$ ($n_\downarrow$) denotes the majority (minority) spin density. It has a (local) maximum for odd integer filling factors and a minimum ($\mathcal{P} = 0\,\%$) for even integer values of $\nu$. In between these two limiting cases, $\mathcal{P}$ changes linearly. The spin polarization at odd integer values of $\nu$ increases for smaller filling factors and equals $\mathcal{P} = 100\,\%$ for $\nu = 1$. However, complete spin polarization can only be achieved in the limit of large Zeeman energies. More precisely, the Zeeman splitting must be larger than the disorder broadening of the Landau levels.

For filling factors $\nu < 1$, where the FQHE is dominant, one may expect the electron system to be always spin polarized, given the argumentation above. In this case, the spin degree of freedom would be frozen out. However, since the Zeeman energy in GaAs is rather weak and has a value comparable to the energy scale of the FQHE, i.e. the effective cyclotron energy of composite fermions, the FQHS are not necessarily fully polarized. This was first pointed out by Halperin in 1983 shortly after the discovery of the FQHE [12]. Since then, various unpolarized and partially polarized FQHS have been found. The question to what extent a FQHS is polarized depends sensitively on the interplay of Zeeman and Coulomb energy. Hence, it is not surprising that transitions between phases of different spin polarization can be induced. A prominent example is the spin transition from unpolarized to fully polarized occurring at filling factor $\nu = 2/3$ [1–3, 13]. Other spin transitions are known to exist, for instance at $\nu = 3/5$ [1], $\nu = 4/3$ [14] and since recently also in the second Landau level at $\nu = 8/3$ [15]. The appearance of spin transitions in the FQHE can be understood intuitively in the composite fermion picture introduced in Sect. 2.4.2. In general, spin transitions occur when CF Landau levels of opposite spin orientation cross. In the FQHE, such crossings can be induced simply by tuning the magnetic field while keeping the filling factor constant since the Zeeman and the Coulomb energy have a different functional dependence on $B$. The separation of CF Landau levels given by the effective cyclotron energy $E_c^*$ depends on the Coulomb energy and is therefore proportional to $1/d = \sqrt{n_e}$, where $d$ is the average particle distance. For a fixed filling factor, this implies $E_c^* \propto \sqrt{B}$. The Zeeman energy in contrast, which splits a single CF Landau level into branches of opposite spin orientation, increases linearly with the magnetic field. We illustrate this in Fig. 3.1 for the case of a spin transition at filling factor $\nu = 2/3$. This transition will play a central role for the remainder of this chapter. In the CF picture, $\nu = 2/3$ corresponds to the case of two completely filled CF Landau levels ($\nu^* = 2$). In the limit of small Zeeman energies, i.e. for low magnetic fields and consequently small densities, the two lowest energy levels have their electron spins aligned in opposite directions. Hence, the electron system is unpolarized. When increasing the magnetic field, the Zeeman splitting grows more rapidly than the cyclotron energy. At some point, a crossing of CF Landau levels occurs. Here, the electron system undergoes a spin transition to an

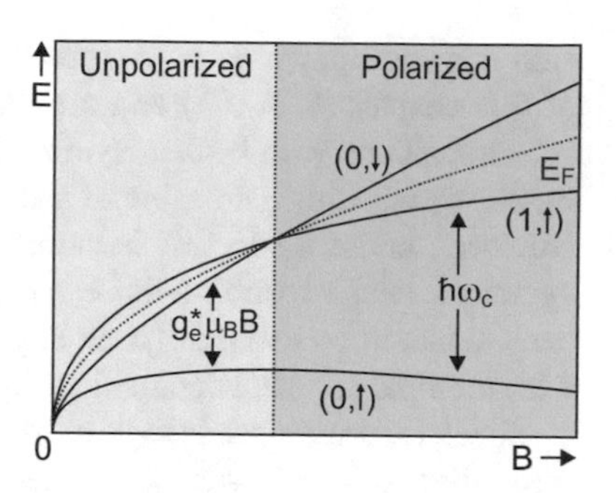


**Fig. 3.1** Evolution of CF Landau levels at $\nu = 2/3$ as a function of the magnetic field strength (based on [11])

unpolarized state. At the crossing point, the energy gap vanishes due to the presence of unoccupied states at the Fermi energy. Consequently, the longitudinal resistance increases, and the quantization of the Hall resistance is lost. Such indications for a spin transition have been observed in transport experiments [1, 3]. Also a direct proof of the transition to a polarized state has been achieved by nuclear magnetic resonance techniques [16] and photoluminescence spectroscopy [2].

Spin transitions in the FQHE can be induced by different means. Besides the method discussed so far, i.e. changing the electron density, an often used technique is tilted-field studies [1, 13, 14]. Tilting the sample with respect to the external magnetic field while keeping the perpendicular field component constant selectively enhances the Zeeman energy. The Coulomb energy, in contrast, only depends on the perpendicular field component and remains therefore unchanged. An alternative method to change the Zeeman energy, though experimentally more difficult, is to apply pressure to the sample [17]. Apart from that, the g-factor can be influenced by the strength of the quantum confinement [18].

The crossing of energy levels with different spin orientations gives rise to complex microscopic spin formations, which bear strong similarities with a ferromagnetic state. This quantum Hall ferromagnetism manifests itself in transport experiments by hysteretic behavior as well as Barkhausen jumps in the sample resistance [3, 5]. Also a coexistence of unpolarized and polarized spin domains has been found by nuclear magnetic resonance spectroscopy [16]. Further insight into the microscopic nature of the spin transitions has been gained using a single-electron transistor on the sample surface to sense the compressibility of the 2DES underneath [11]. With this technique, domain sizes in excess of 500 nm were observed. More recently, real-space imaging of the spin transition at $\nu = 2/3$ has been achieved by scanning optical microscopy using trions (charged excitons) as local probes for the electron spin polarization [6]. These experiments provided an intriguing visualization of the spin transition and revealed domain sizes of about 33 μm.

Quantum Hall ferromagnetism is not unique to the FQHE but also appears in the IQHE [19]. The most prominent example is the IQHE at $\nu = 1$. Here, a fully aligned spin system is favored by the electron-electron interaction even in the case of vanishing Zeeman energy [7, 20]. It can be considered a Heisenberg-like isotropic ferromagnet [20]. Interestingly, the $\nu = 1$ quantum Hall state is also home to complex spin textures known as skyrmions [7, 21]. These will be important in a later part of this chapter (Sect. 3.3) and are discussed there in more detail. Also Ising-type spin transitions with broken symmetry can be realized in the IQHE similar to the FQHE by inducing a crossing of Landau levels with opposite spin orientations [22]. However, in the IQHE this is hampered by the fact that both the Zeeman and the cyclotron energy exhibit a linear $B$-dependence. Nevertheless, spin transitions have been achieved by using large tilt angles [22] or by resorting to wide quantum wells with multiple subband occupation [4].

Special attention needs to be attributed to the spin physics at filling factor $\nu = 1/2$. According to the CF theory, the electron system at $\nu = 1/2$ can be understood as a Fermi sea of weakly interacting composite fermions in a vanishing effective magnetic field. Yet, it is important to bear in mind that the effective magnetic field only applies to the orbital degree of the composite fermions. The spin degree of freedom is subject to the total magnetic field. In the case of non-interacting particles, composite fermions have a parabolic energy dispersion $E = \frac{\hbar^2 k_{CF}^2}{2m_{CF}^*}$, with $m_{CF}^*$ being the effective mass of composite fermions and $k_{CF}$ their wavevector. Hence, the density of states for each spin orientation is constant as shown in Fig. 3.2. The overall spin polarization depends on the relative occupation of the two spin branches. For $E_z > E_F$, the electron system is fully polarized. Below this point, it becomes partially polarized. The spin polarization can therefore be continuously tuned by changing the density $n_e$ and thereby the magnetic field while keeping $\nu = 1/2$ constant. This has been confirmed experimentally by polarization-sensitive photoluminescence spectroscopy [2]. The transition to a fully polarized electron system occurred around 9 T. The above considerations are similarly applicable to the CF Fermi sea at the particle-hole-conjugate filling factor $\nu = 3/2$ and to higher-order CF Fermi seas, e.g. at $\nu = 1/4$, where four flux quanta are attached to each electron.

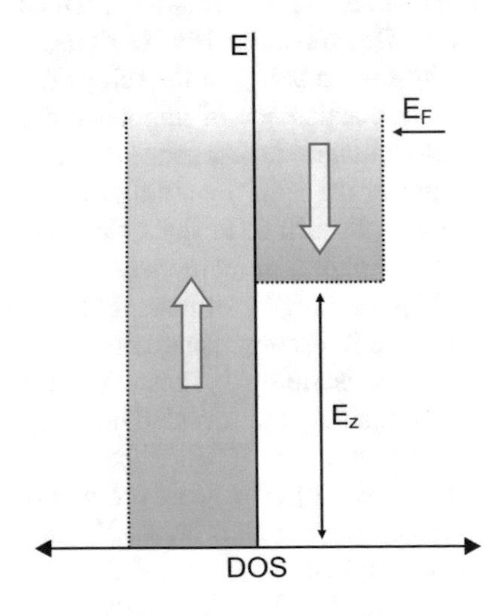


**Fig. 3.2** Relative occupation of the spin-*up* and -*down* energy levels for a CF Fermi sea forming at high magnetic fields (based on [23])

### 3.1.2 The Nuclear Spin System in a GaAs/AlGaAs Heterostructure

When a single nuclear spin is placed in a magnetic field, it experiences in analogy to its electronic counterpart the Zeeman interaction

$$E_z = -\boldsymbol{\mu}_M \cdot \boldsymbol{B}, \tag{3.2}$$

where $\boldsymbol{\mu}_M$ is the nuclear magnetic moment defined as

$$\boldsymbol{\mu}_M = \frac{g_N \mu_N}{\hbar} \boldsymbol{J}. \tag{3.3}$$

In this equation, $g_N$ denotes the nuclear g-factor, $\mu_N$ the nuclear magneton $\mu_N = \frac{e\hbar}{2m_p}$ with the proton rest mass $m_p$ and $\boldsymbol{J}$ is the nuclear spin angular momentum. Quantum mechanics dictates that the projection of $\boldsymbol{J}$ along the magnetic field direction has discrete values $\hbar m_J$, where the magnetic quantum number $m_J$ is restricted to values $\mathcal{J}, \mathcal{J}-1, \ldots, -\mathcal{J}+1, -\mathcal{J}$. Here, $\mathcal{J}$ is the nuclear spin quantum number. Altogether, Eq. 3.2 is transformed into

$$E_z = -g_N \mu_N B m_J. \tag{3.4}$$

An often used quantity in this context is the gyromagnetic ratio $\gamma$. It relates the transition frequency of the Zeeman splitting $f$ directly to the magnetic field via $f = \frac{\gamma}{2\pi} B$.

The magnetic moments and gyromagnetic ratios of the nuclear isotopes inherent to GaAs are shown in Table 3.1 together with other properties which will become important later on. Since the electron wavefunction is located almost entirely inside of the GaAs quantum well, the influence of the Al nuclei in the barrier can be neglected in most cases.

All stable nuclei in GaAs have spin $\mathcal{J} = 3/2$. Hence, when applying a magnetic field, the nuclear spin system splits equally into four energy levels due to the Zeeman energy in expression 3.4 (Fig. 3.3a). Each of these levels corresponds to a different spin orientation. Consequently, the relative occupation of the energy levels

**Table 3.1** Properties of the stable nuclear isotopes in GaAs (data taken from [10, 24])

| Isotope | $^{69}$Ga | $^{71}$Ga | $^{75}$As |
|---|---|---|---|
| Abundance (%) | 60.1 | 39.9 | 100 |
| Spin quantum number $\mathcal{J}$ | 3/2 | 3/2 | 3/2 |
| Magnetic moment $\mu_M$ ($\mu_N$) | 2.0166 | 2.5623 | 1.4395 |
| Reduced gyromagnetic ratio $\frac{\gamma}{2\pi}$ (MHz/T) | 10.248 | 13.021 | 7.3150 |
| Hyperfine constant $A_{HF}$ (μeV) | 38 | 49 | 46 |
| Overhauser field at full polarization $B_N$ (T) | −1.37 | −1.17 | −2.76 |

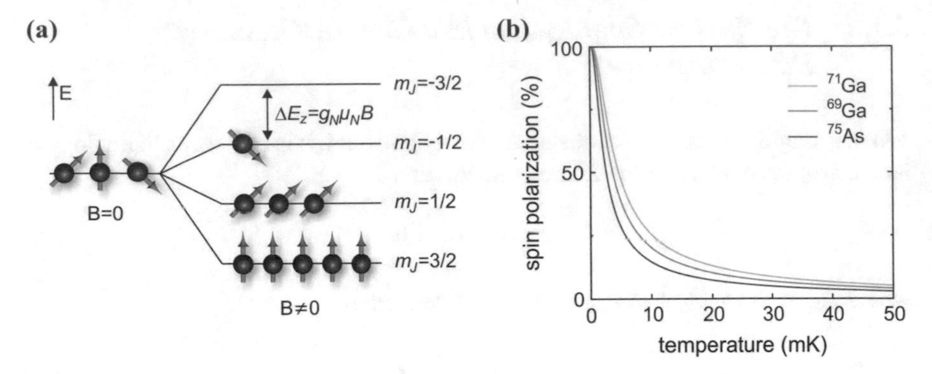


**Fig. 3.3** **a** Zeeman splitting of a nuclear spin system with spin $\mathcal{J} = 3/2$. **b** Nuclear spin polarization of the nuclides $^{69}$Ga, $^{71}$Ga and $^{75}$As as a function of temperature at a magnetic field of 5 T. Calculations are based on Eq. 3.5

determines the overall nuclear spin polarization. If the nuclei are equally distributed among these levels, the nuclear spin system is unpolarized. This corresponds to the situation at high temperatures. When lowering the temperature, the number of nuclei in the lowest spin level increases and with it the nuclear spin polarization. Based on Boltzmann statistics, the nuclear spin polarization $\mathcal{P}_N$ can be calculated according to

$$\mathcal{P}_N = B_{\mathcal{J}}(\eta) = \frac{2\mathcal{J}+1}{2\mathcal{J}} \coth\left(\eta \frac{2\mathcal{J}+1}{2\mathcal{J}}\right) - \frac{1}{2\mathcal{J}} \coth\left(\eta \frac{1}{2\mathcal{J}}\right). \tag{3.5}$$

Here, the Brillouin function $B_{\mathcal{J}}$ and the substitution $\eta = g_N \mu_N \mathcal{J} B / k_B T$ was used, with $k_B$ representing the Boltzmann constant and $T$ the temperature of the nuclear spin system [10]. Figure 3.3b shows exemplarily the temperature dependence of $\mathcal{P}_N$ for the three isotopes $^{69}$Ga, $^{71}$Ga and $^{75}$As at a magnetic field of 5 T. It emphasizes that the nuclear spin polarization diminishes rapidly with increasing temperature. If a high spin polarization is desired in thermal equilibrium, low temperatures as well as high magnetic fields are required. Many of the experiments described in this thesis were performed inside of a dilution refrigerator at a base temperature of around 20 mK, which corresponds to a nuclear spin polarization of roughly 12 % at 5 T.

### *3.1.3 The Hyperfine Coupling of Electronic and Nuclear Spins*

Treating the nuclear spins as an isolated system of paramagnets as done in the previous section does not grasp the full complexity of the nuclear spin system. In a more realistic scenario, nuclear spins interact with their environment, for instance via the dipole-dipole interaction. Another important mechanism of nuclear spin interaction is the hyperfine coupling to the electron spin system of the 2DES. For a

GaAs-based 2DES, the hyperfine interaction is dominated by the Fermi contact interaction between the s-type conduction band electrons and the nuclei [10]. This interaction can be described by the Hamiltonian

$$H_{HF} = A_{HF} \boldsymbol{J} \cdot \boldsymbol{S}, \tag{3.6}$$

with the hyperfine interaction constant

$$A_{HF} = \frac{2\mu_0}{3}(g_e\mu_B)(g_N\mu_N)\,|\psi(0)|^2\,, \tag{3.7}$$

which depends on the probability $|\psi(0)|^2$ of finding an electron at a nucleus' site [16]. In the case of s-band electrons, the overlap of the electron wavefunction with the nucleus is strong, and $|\psi(0)|^2$ has a large value. The hyperfine constants for GaAs are stated in Table 3.1.

The product $\boldsymbol{J} \cdot \boldsymbol{S}$ in Eq. 3.6 can be rewritten as

$$\boldsymbol{J} \cdot \boldsymbol{S} = \frac{1}{2}\left(J^+S^- + J^-S^+\right) + J_zS_z, \tag{3.8}$$

where the superscript + (−) denotes the raising (lowering) of the respective spin operators and the subscript $z$ represents the z-component of the spins.

The first term is called the "flip-flop" term. It describes a nuclear spin flip mediated by an electron spin rotation in the opposite direction, i.e. a spin transfer from an electron to the nuclear spin system and vice versa. This dynamic term is key to different ways of dynamic nuclear spin polarization, for example by electron spin resonance [25] and optical pumping [26, 27]. It also provides an efficient channel for nuclear spin relaxation if the energies of electronic and nuclear spin flip match [28, 29].

The second, static term in Eq. 3.8 states that a polarized nuclear spin system modifies the electronic Zeeman energy. This contribution to $E_z$ is often quantified by an additional effective magnetic field $B_N$ acting solely on the electronic spin degree of freedom, i.e. the filling factor remains unchanged. The change of the electronic Zeeman energy caused by this so-called Overhauser field can be observed in electron spin resonance experiments [25, 30]. Since the hyperfine interaction selectively alters the Zeeman energy, it also affects the occurrence of spin transitions in the FQHE. Some values of $B_N$ calculated for the case of fully polarized nuclear spin system are listed in Table 3.1. The negative sign of $B_N$ indicates that the Overhauser field is oriented opposite to the external magnetic field. Thus, increasing the nuclear spin polarization leads to a reduction of the electronic Zeeman energy. Analogously to the Overhauser field, also a spin-polarized electron system causes a change of the nuclear Zeeman energy. This effect is detectable in nuclear magnetic resonance experiments as a shift of the resonance frequency—also known as the Knight shift [31]. Measurements of the Knight shift have been proven a powerful method to probe the electron spin polarization [32, 33] and their spatial variation [16]. This technique will become important in Chaps. 4 and 5.

## 3.2 Nuclear Spin Relaxation Rate in the Quantum Hall Regime

The flip-flop mechanism for nuclear spin relaxation introduced in the previous section requires the energy gain of a nuclear spin flip to be compensated by the electron system, for example by an increase of its kinetic energy. In a metal, this energy conservation is easily accomplished since a continuum of states is available for either spin direction. In the quantum Hall regime the situation is different. Here, the DOS is bundled in Landau levels, and states of opposite spin orientation are separated by the electronic Zeeman energy. This splitting is about three orders of magnitude larger than the nuclear spin splitting. Hence, in the quantum Hall regime the energy conservation is in general not easily fulfilled, and the nuclear spin relaxation via this channel is rather slow. However, at certain filling factors low-energy spin excitations exist in the 2DES, which provide an efficient way of nuclear spin relaxation. This section aims at identifying these low-energetic spin excitations by measuring the nuclear spin relaxation rate at different filling factors. Of particular interest is here the nuclear spin relaxation at the enigmatic $\nu = 5/2$ FQHS. Theory predicts the existence of skyrmionic spin excitations in its close vicinity [34]. If present, signatures of these excitations would be observable in the nuclear spin relaxation rate. Before looking at the experimental results, we first introduce the technique used for measuring the relaxation rate. It is based on the spin transition at $\nu = 2/3$.

### *3.2.1 Nuclear Magnetometry Based on the $\nu = 2/3$ Spin Transition*

In order to measure the nuclear spin relaxation rate, two tasks need to be accomplished. First, the nuclear spin polarization must be driven out of thermal equilibrium. In a second step, the time dependent relaxation of the nuclear spin polarization back to its equilibrium value needs to be measured. Thus, two tools are necessary—one for manipulating and one for measuring the nuclear spin polarization. Both of these tasks can be accomplished by taking advantage of the spin transition at $\nu = 2/3$ as described below.

**Nuclear Spin Manipulation**

The spin transition at $\nu = 2/3$ is well suited to manipulate the nuclear spin polarization (Sect. 3.1.1). The crossing of Landau levels with opposite spin orientation allows electronic spin flips at little energetic cost, thus, matching the nuclear Zeeman splitting more aptly. This opens up an efficient channel for conveying spin angular momentum between both systems. Based on this situation, nuclear spin manipulation can be realized simply by driving a strong current through the sample. Imposing a strong current has two main consequences. On the one hand, it inevitably causes a heating of the electron system. This heat will be transferred to the nuclear spin system, which leads according to Boltzmann statistics to a lower spin polarization (Fig. 3.3b). On the other hand, the coexistence of unpolarized and polarized domains at the spin transition was found to facilitate a dynamic nuclear spin polarization when imposing an external current [3, 28, 29, 35, 36]. The underlying process can be understood in the following way [10]. If electrons are driven across the domain walls separating regions of opposite spin orientation, a spin reversal is required. This spin flip may be carried out by a flip-flop process involving a simultaneous nuclear spin flip in the opposite direction. Thus, current flow may locally enhance the nuclear spin polarization in a dynamic fashion. The altered spin polarization acts back on the electronic domain morphology because it locally changes the electronic Zeeman energy via the hyperfine interaction. This implies spatial fluctuations of the electronic Zeeman energy across the sample, which presumably leads to smaller domain sizes and therefore higher electron scattering [36]. Indeed, the longitudinal resistance at the spin transitions was found to be enhanced considerably under the influence of a strong current [35, 36]. In transport experiments a large resistance peak forms over a long time scale, which indicates the involvement of nuclear spins. More direct evidence for a current-induced nuclear spin polarization is given by the appearance of strong nuclear magnetic resonance signals in the region of the spin transition [3]. Nevertheless, a detailed microscopic picture of the underlying physics is missing up to date.

Which of these two counteracting effects of the current flow is strongest presumably depends on the exact details of the experiment, e.g. current strength and sample disorder. We will come back to this point in the next section. A measurement of the nuclear spin relaxation rate is possible in either case as long as the system is driven out of equilibrium in the first place.

**Detection of the Nuclear Spin Polarization**

Changes of the nuclear spin polarization can be probed simply by measuring a displacement of the spin transition at $\nu = 2/3$. As mentioned earlier, the coincidence condition is set by the ratio between Coulomb and Zeeman energy. The latter depends on the nuclear spin polarization due to the hyperfine interaction. Yet, the question remains how to detect the position of the spin transition. This can be done easily by standard transport experiments as shown in Fig. 3.4. Here, the longitudinal resistance was measured around $\nu = 2/3$ for different densities and consequently also different magnetic fields. The spin transition is clearly visible by a peak of finite resistance separating regions of vanishing $R_{xx}$. The resistance peak indicates the closing of the energy gap induced by the swapping of the Landau level spin branches. Below this point the electron system is unpolarized, whereas at higher magnetic fields it is completely spin polarized.

**Putting it all Together**

The ability to manipulate and measure the nuclear spin polarization provides us with the basic tools to determine the nuclear spin relaxation rate. The entire measurement sequence is shown in Fig. 3.5. It exploits the varying degree of electron-nuclear spin interaction at different filling factors. Rapid changes of the filling factor were

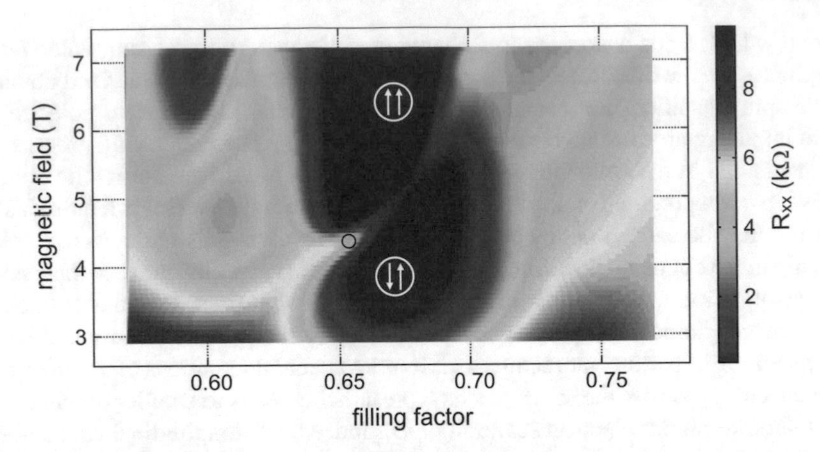


**Fig. 3.4** Magnetic field dependence of the longitudinal resistance around filling factor $\nu = 2/3$. The crossing of CF Landau levels manifests itself as a peak of finite resistance traversing the region of $R_{xx} = 0$. At lower magnetic fields the electron system is unpolarized; at larger fields it is polarized. The *black circle* marks the filling factor used for the experiments in Fig. 3.6

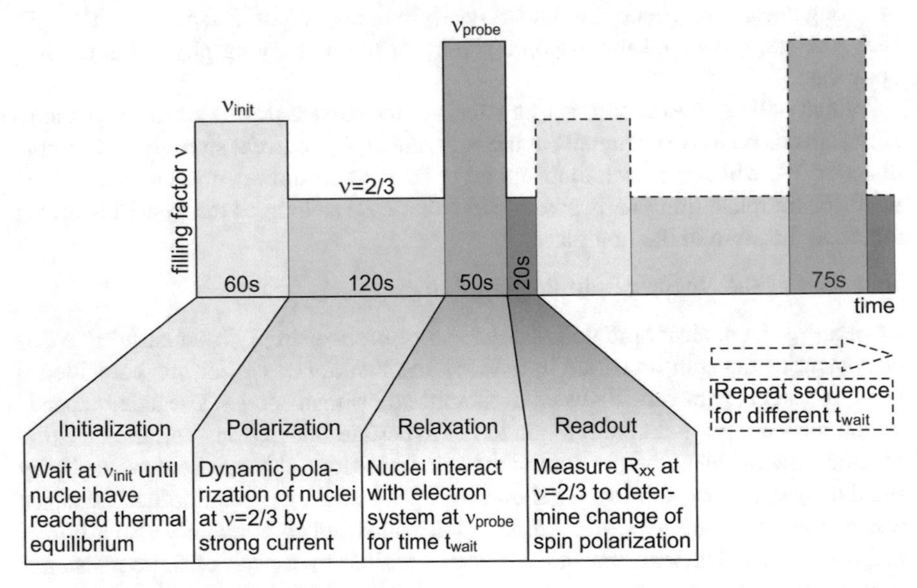


**Fig. 3.5** Measurement sequence used to determine the nuclear spin relaxation rate at filling factor $\nu_{\text{probe}}$ (see text for details)

achieved by tuning the electron density electrostatically with the help of a backgate. The magnetic field was kept constant. The measurement sequence consists basically of four steps. In the *first step*, the nuclear spin system is initialized in order to have a common starting point for all subsequent steps. This is done by completely thermalizing the nuclear spin system to the environment. For this purpose, the electron system is set to a filling factor $\nu_{\text{init}}$ which provides a fast nuclear spin relaxation. We have chosen $\nu_{\text{init}}$ to be in the vicinity of $\nu = 1$ out of reasons which will become apparent in the next section. During the *second step*, the nuclei get dynamically polarized by applying a strong current (200 nA) at $\nu = 2/3$. In the *third step*, the electron density is tuned to the filling factor of interest $\nu_{\text{probe}}$ where the nuclear spin relaxation rate is measured. Therefore, the nuclear spin system is allowed to interact with the electron system for the time $t_{\text{wait}}$. In the *fourth step*, the 2DES is returned to $\nu = 2/3$ for detecting the change of nuclear spin polarization which has occurred during $t_{\text{wait}}$. This is done simply by measuring the longitudinal resistance at the flank of the spin transition peak. These four steps are repeated multiple times while for each cycle a different value of $t_{\text{wait}}$ is chosen in order to obtain the whole relaxation curve.

An example of nuclear spin relaxation is depicted in Fig. 3.6a. After fitting an exponential function $R = R_0 + \Delta R \cdot e^{-\frac{t}{\tau}}$, the relaxation time $\tau$ can be extracted. It is important to emphasize that $\tau$ is only equal to the nuclear spin relaxation time $T_1$ if the resistance change $\Delta R_{xx}$ depends linearly on the electronic Zeeman energy. This is approximately fulfilled for small changes of $E_z$ [23]. In fact, the exponential behavior of $R_{xx}$ in Fig. 3.6a supports the interpretation of $\tau$ as the nuclear spin relaxation time. Apart from that, even if $R_{xx}$ responds to changes in $E_z$ in a non-linear way, $\tau$ remains a good measure to compare relaxation rates at different filling factors and to identify filling factors of fast nuclear spin relaxation. The exact point selected for dynamic polarization and read-out of the nuclear spin polarization is indicated in Fig. 3.4. It was chosen for its large responsiveness to an external current. Fig. 3.6b shows the effect of a strong current (200 nA) on the longitudinal resistance, recorded over an hour. At first, $R_{xx}$ increases strongly and starts to saturate after about 20 min. A short excursion to a different filling factor causes a drop in $R_{xx}$ due to the partial relaxation of the nuclear spins. This observation highlights the basic principle underlying the measurement scheme in Fig. 3.5.

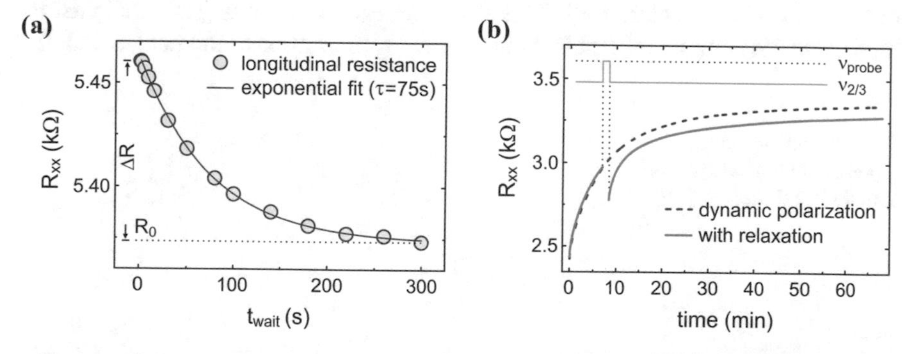


**Fig. 3.6** **a** Measurement of the nuclear spin relaxation rate via changes in the longitudinal resistance. The spin transition at $\nu = 2/3$ was used for the manipulation and detection of the nuclear spin polarization. A nuclear spin relaxation time of 75 s was extracted from an exponential fit. **b** Time dependence of the longitudinal resistance at $\nu = 2/3$ under the influence of a strong current (200 nA). An intermediate relaxation of the nuclear spin polarization causes a drop of $R_{xx}$

### 3.2.2 Results and Discussion

Using the previously introduced technique, we measured the nuclear spin relaxation rate at all filling factors within the tuning range of the backgate. The measurements were performed at a constant magnetic field of 4.3 T and a temperature of roughly 20 mK. Two sets of experiments were carried out. First, a low-frequency current ($\sim$10 Hz) of 200 nA was driven constantly through the sample during each step of the measurement sequence. The main effect of the strong current on the measurement is a substantial heat input and consequently a higher electron temperature. In a second set of experiments, the current was turned off during nuclear spin relaxation at $\nu_{\text{probe}}$ and was applied only for the dynamic polarization as well as the read-out of the nuclear spin polarization.

The filling factor dependence of the nuclear spin relaxation rate obtained in the first case is shown in Fig. 3.8a together with the longitudinal resistance for these conditions. The salient observation is two regions of fast nuclear spin relaxation symmetric around filling factor $\nu = 1$. This is the region used for the initialization of the nuclear spin system ($\nu_{\text{init}}$ in Fig. 3.5). A similar relaxation characteristic has been observed before at $\nu = 1$ [28, 29, 38]. It was attributed to the formation of skyrmions in the vicinity of the $\nu = 1$ IQHS, more precisely, to the presence of a so-called Goldstone mode. Skyrmions are gapped collective spin excitations with charge $\pm e$ [7]. They possess a complex spin texture which can be illustrated as a vortex of spins (Fig. 3.7). In the center of the vortex, the electron spin is directed opposite to the external magnetic field. When going towards the perimeter, the spin slowly rotates by 180° and ends up oriented along the magnetic field. This spin order results from a competition between Zeeman and exchange energy. The latter prefers to distribute a single spin flip among many electrons instead of reversing a single electron spin. In fact, the optimal skyrmion configuration requires only half the energy of a single electron spin flip [7, 19]. The size and effective spin of the skyrmions depends on the ratio of Zeeman and Coulomb energy. At $T = 0$, the skyrmions are predicted to order in a crystal structure [39]. This skyrmion crystal exhibits gapless Goldstone modes, which open up an efficient channel for spin transfer to the nuclei. The fast

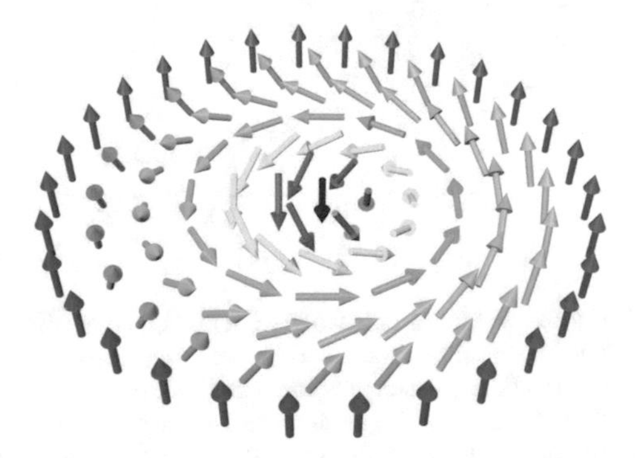

**Fig. 3.7** Schematic representation of a skyrmion spin texture (based on [37])

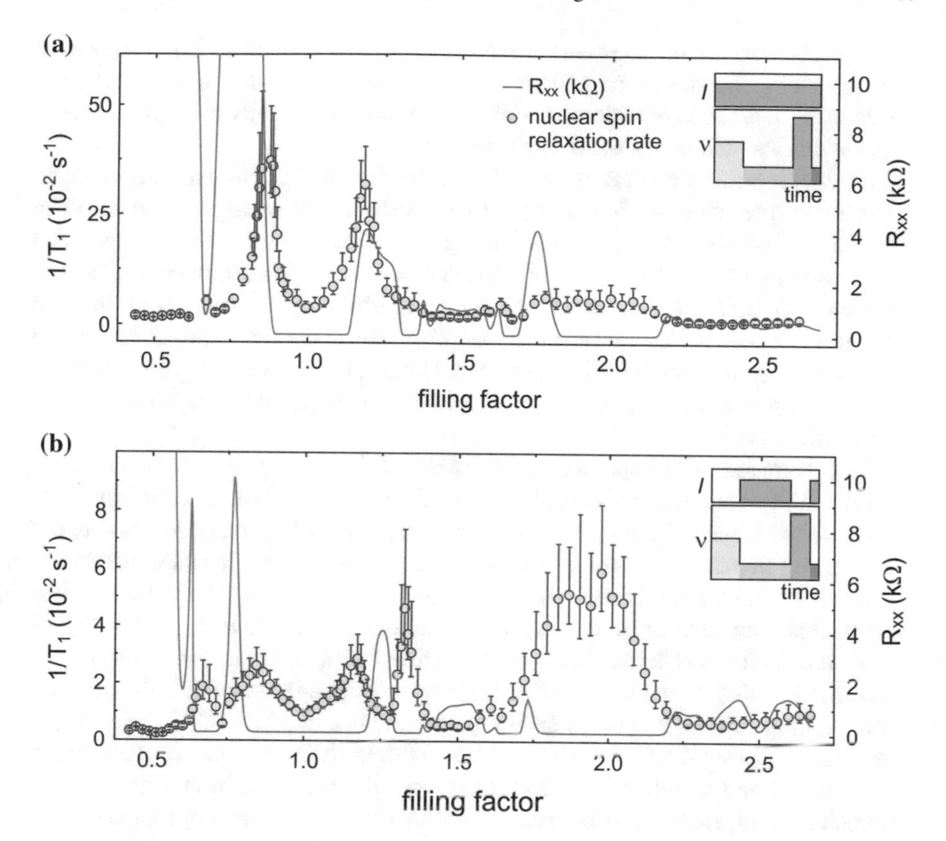


**Fig. 3.8** **a** Filling factor dependence of the nuclear spin relaxation rate. A strong current of 200 nA was driven constantly through the sample. The inset shows the measurement sequence according to Fig. 3.5. **b** Measurement of the nuclear spin relaxation rate while the current is turned off during relaxation

nuclear spin relaxation in Fig. 3.8a is a consequence of these Goldstone modes [40]. A strong coupling between the electronic and nuclear spin system around $\nu = 1$ was also inferred from heat capacity measurements [41]. The existence of skyrmions was further corroborated by the observation of a strongly decreasing electron spin polarization when going away from exact $\nu = 1$ [32, 42]. Apart from the $\nu = 1$ quantum Hall state, skyrmions are also expected to occur at other (odd) integer and fractional quantum Hall states with similar energetic situations, for instance at $\nu = {}^1\!/_3$ and $\nu = {}^1\!/_5$ [7]. However, experimental evidence for this assumption is scarce. Recently, the existence of skyrmions has been predicted by theory for the $\nu = {}^5\!/_2$ state as well [34].

When turning off the external current during nuclear spin relaxation, the electron temperature is lowered substantially. Under such conditions, additional features appear in the filling factor dependence of $1/T_1$ as shown in Fig. 3.8b. These are addressed hereinafter starting at low filling factors:

- The first relaxation peak appears around $\nu = 2/3$. The fast nuclear spin relaxation at this filling factor can be understood as a consequence of the spin transition occurring here. The coincidence of CF Landau levels used to dynamically polarize the nuclei also facilitates a rapid spin relaxation.
- The double peak structure around $\nu = 1$, which was previously attributed to skyrmion formation, is also present at base temperature, though much lower in intensity. The observation of a decreasing spin relaxation at lower temperatures is consistent with the theoretical prediction of a Korringa-like temperature dependence $1/T_1 \propto T$ [40]. A similar behavior was observed by Tracy et al. [43]. In contrast, Gervais et al. found an increased relaxation in the same temperature range upon cooling the system [44]. For small deviations from $\nu = 1$, Fig. 3.8b displays a linear dependence $1/T_1 \propto |1 - \nu|$. This behavior nicely confirms the theoretical predictions [40].
- The next region of fast spin relaxation occurs at the $\nu = 4/3$ quantum Hall state. In the limit of small Zeeman energy, $\nu = 4/3$ is the particle-hole conjugate state to $\nu = 2/3$. Hence, the $\nu = 4/3$ state also undergoes a spin transition in roughly the same magnetic field range. As in the case of the $\nu = 2/3$ state, this would account for the high relaxation rate. This interpretation is supported by the slow spin relaxation occuring at $\nu = 5/3$. As the conjugate state to $\nu = 1/3$, this FQHS does not exhibit a spin transition (only one filled CF Landau level). Interestingly, also the FQHS at $\nu = 8/5$ has a slightly increased relaxation rate. It is the particle-hole conjugate state to $\nu = 2/5$ and consequently has two filled CF Landau levels as well. However, the $8/5$ state is subject to a different Coulomb energy compared to $\nu = 4/3$ and therefore is further apart from the spin transition at the present magnetic field. Hence, it is not surprising that the enhancement of the relaxation rate is weaker here.
- Rather surprising is the observation of a fast nuclear spin relaxation around $\nu = 2$. To our knowledge, this is the first report of such a behavior. No other measurements of the relaxation rate are known to exist in this filling factor range at such low temperatures. So far, no clear explanation has been found for the observed behavior. In the single-particle picture the excitation gap at $\nu = 2$ is set by the cyclotron energy and has a large value. Hence, excitations between different Landau levels fail to explain the fast nuclear spin relaxation. The exchange energy, being a dominant factor at $\nu = 1$, is much smaller around $\nu = 2$, which renders the formation of skyrmions unlikely. Possibly, edge effects are responsible for the fast relaxation. However, in this case, the relaxation curve would show signatures of two distinct relaxation times, one for the bulk and one for edge effects. We also verified that the observed behavior is not an artifact of the detection method by using filling factor $\nu = 1/2$ as an alternative detection scheme [23]. Our findings suggest that the physics at $\nu = 2$ is in fact richer than expected. The microscopic origin of these results remains to be resolved.
- An important motivation for these experiments was the question whether signatures of the $\nu = 5/2$ state would show up in the nuclear spin relaxation rate. As mentioned earlier, theory predicts the formation of skyrmions for this FQHS [34]. Evidently, no indication of an enhanced relaxation rate is observed in Fig. 3.8b. Also a more

detailed measurement around $\nu = {}^5\!/_2$ unveiled no further insights. Possibly, the quality of the $\nu = {}^5\!/_2$ state is not sufficient for the formation of skyrmions. On the other hand, this result may indicate that the theoretical understanding of the $\nu = {}^5\!/_2$ state is not accurate enough.

Figure 3.8 demonstrates that measurements of the nuclear spin relaxation rate can be utilized to probe low-energetic spin excitations in the integer and fractional quantum Hall effect. Our results highlight the rich spin physics present in the quantum Hall regime. Another interesting facet of spin physics is the filling factor dependence of the nuclear spin polarization studied in the following section.

## 3.3 Filling Factor Dependence of the Nuclear Spin Polarization

When taking a closer look at the measurement data acquired for Fig. 3.8, it becomes apparent that not only the relaxation rate depends on the filling factor but also the resistance change ($\Delta R$ in Fig. 3.6). In other words, when waiting for a long time at $\nu_{\text{probe}}$, the resistance value $R_0$ measured upon return to $\nu \approx {}^2\!/_3$ depends on the exact value of $\nu_{\text{probe}}$ as shown in Fig. 3.9. This implies that the nuclear spin polarization $\mathcal{P}_N$ in thermal equilibrium changes for different filling factors if our interpretation of $R_{xx}(\nu = {}^2\!/_3)$ as a measure of the nuclear spin polarization is correct. First indications of such a behavior have been observed previously [45].

In order to further investigate this hypothesis, we measured directly the shift of the spin transition according to the scheme in Fig. 3.10a. In the first step, the electron system rests at filling factor $\nu_{\text{probe}}$ for 10 min to ensure thermal equilibrium. During this time, the current is switched off. In a second step, the equilibrium nuclear spin polarization is measured by quickly jumping to $\nu \approx {}^2\!/_3$ and sweeping across the spin transition peak while keeping the magnetic field constant at 5 T. This process is repeated afterwards for different values of $\nu_{\text{probe}}$. The result is shown in Fig. 3.10b as

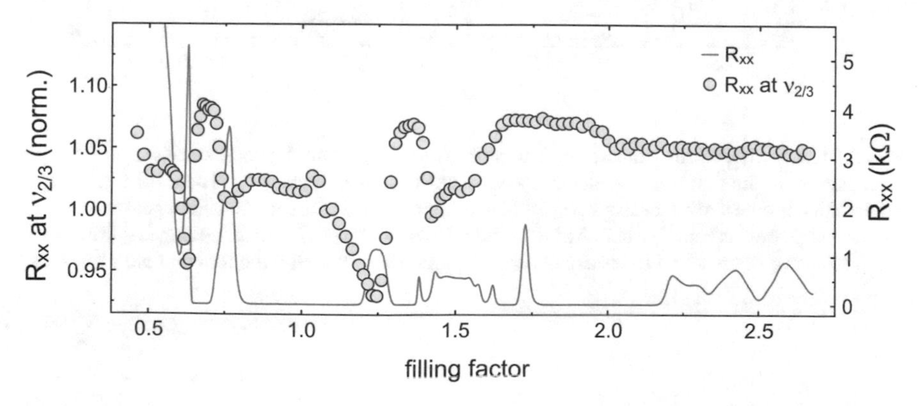


**Fig. 3.9** Longitudinal resistance at $\nu = {}^2\!/_3$ measured after the nuclei have equilibrated at $\nu_{\text{probe}}$. The resistance values at $\nu = {}^2\!/_3$ are normalized by the values obtained after dynamic nuclear polarization

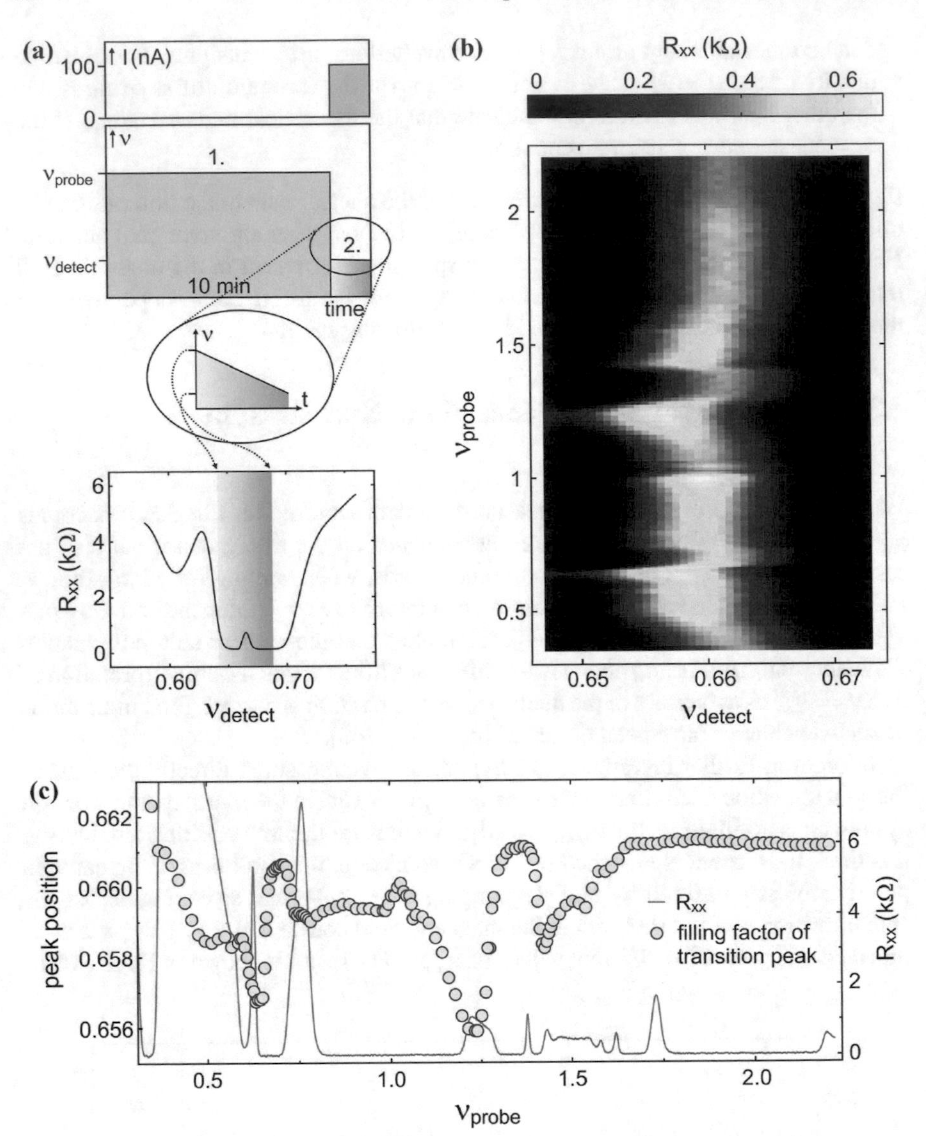


**Fig. 3.10** **a** Measurement sequence used to probe the filling factor dependence of the nuclear spin polarization. In the first step, the electron system rests for 10 min at $\nu_{\text{probe}}$. Then, the nuclear spin polarization is detected by measuring $R_{xx}$ while sweeping across the spin transition peak at $\nu = 2/3$. The magnetic field is kept constant. **b** Shift of the spin transition peak in color coding acquired at 5 T. **c** Filling factor of the transition peaks in panel **b** plotted together with the longitudinal resistance

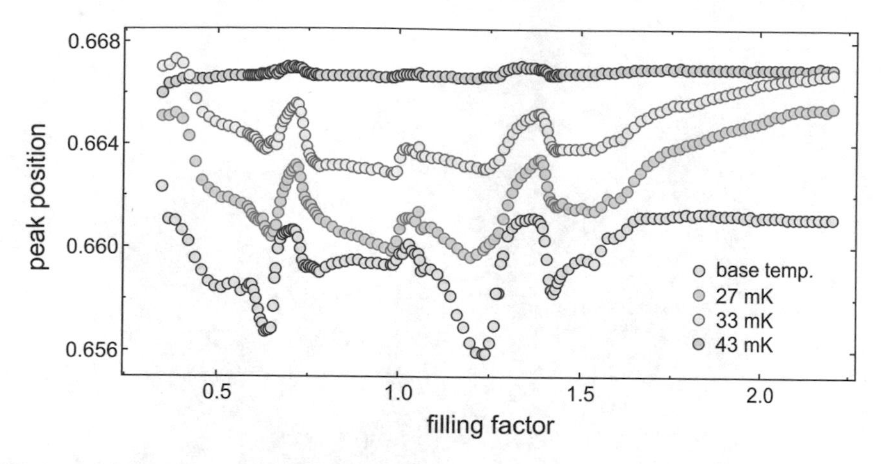


**Fig. 3.11** Filling factor dependence of the nuclear spin polarization measured at different temperatures. Plotted is the peak position (filling factor) of the spin transition at $\nu = 2/3$ ($\propto 1/\mathcal{P}_N$)

a color plot. The extracted peak positions are depicted in Fig. 3.10c together with the longitudinal resistance measured at the same magnetic field. From the comparison with Fig. 3.9, it can be stated that it is indeed a displacement of the spin transition causing the resistance changes in Fig. 3.9. This supports with hindsight the interpretation $R_{xx} \propto \mathcal{P}_N$ used in the previous section to determine the nuclear spin relaxation rate. From the direction of the peak shift, the polarity of the change in $\mathcal{P}_N$ can be assigned. A shift to lower filling factors corresponds to an increase of the nuclear spin polarization. Hence, the nuclear spin polarization is highest around $\nu = 1.22$. At this point, a low resistance was measured in Fig. 3.9. With this correlation in mind, we can clarify the effect of the current driven through the sample for dynamic polarization in the previous section. The steady increase of $R_{xx}$ shown in Fig. 3.6b identifies $\mathcal{P}_N$ to be overall reduced by the current flow.

Not only the position of the spin transition peak shifts for different filling factors $\nu_{\text{probe}}$ but also its height varies. Since the magnitude of the resistance peak is inherently connected to the size and structure of the domain pattern at the $\nu = 2/3$ spin transition, this observation points to an additional source of disorder imposed at certain filling factors $\nu_{\text{probe}}$. Remarkably, this is also the case for the exact filling $\nu_{\text{probe}} = 1$.

Figure 3.11 reveals the temperature dependence of the nuclear spin polarization. The overall shift of the peak position to larger filling factors upon warming the sample results from the population of higher-energy spin levels and the concomitant depolarization of the nuclei. At the same time, the filling factor dependence of $\mathcal{P}_N$ is weakened at higher temperatures. At 43 mK, $\mathcal{P}_N$ becomes independent of the filling factor. Interestingly, not all features disappear uniformly. At first, the increased nuclear spin polarization at $\nu = 1.22$ vanishes, while other features remain unaffected by the temperature change. Figure 3.11 allows quantifying the variation of $\mathcal{P}_N$ which occurs at base temperature. $\Delta\mathcal{P}_N$ is roughly equivalent to the change in nuclear spin polarization caused by a cooling of the electron system from 43 mK to base

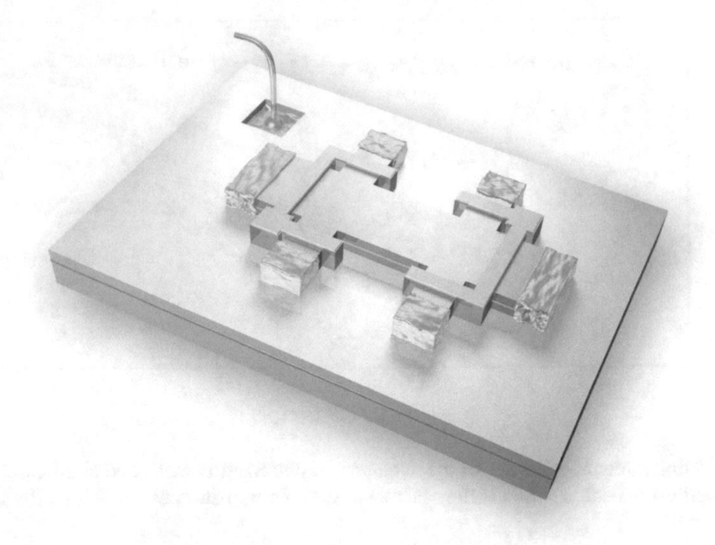

**Fig. 3.12** Schematic rendering of a sample with topgates used to deplete the 2DES underneath and thereby separate the interior of the sample from the contacts. An insulating layer (shown in *blue*) was fabricated below the metallic topgates to eliminate leakage currents

temperature ($\sim$20 mK). With the help of Eq. 3.5, this temperature difference can be converted to $\Delta\mathcal{P}_N = 5\,\%$ for $^{75}$As nuclei.

The observation of a filling-factor-dependent nuclear spin polarization is surprising since $\mathcal{P}_N$, in the first place, depends only on the nuclear Zeeman splitting and the temperature. It would stand to reason to suspect the hyperfine interaction as the origin of the variations in $\mathcal{P}_N$. As mentioned earlier, a spin-polarized electron system reduces the Zeeman splitting of the nuclei (Knight shift). However, this effect is about three orders of magnitude smaller than our findings. A plausible alternative explanation is a varying electron temperature caused by unintended current flows. The sample couples to the environment outside of the cryostat via the leads and can thereby easily pick up microwave radiation from the environment. This would lead to a persistent, small current flow through the sample even when all intentionally imposed external currents are switched off. As a consequence, the 2DES would heat up slightly. The resulting temperature gain would depend on the sample impedance and would therefore change for different filling factors. This scenario would explain why the electron temperature varies and with it also the nuclear spin polarization.

To verify or rule out this hypothesis, we repeated the experiment with the modified sample sketched in Fig. 3.12. It has metallic gates on top of the surface designed such that the 2DES underneath can be depleted locally, thereby isolating the interior part of the sample from the contacts. With this capability at hand, a potential influence of the leads on the equilibrium value of $\mathcal{P}_N$ can be investigated. We have repeated the measurements of Fig. 3.10c with the topgates being energized when approaching thermal equilibrium and turned off for the detection of the spin transition peak.

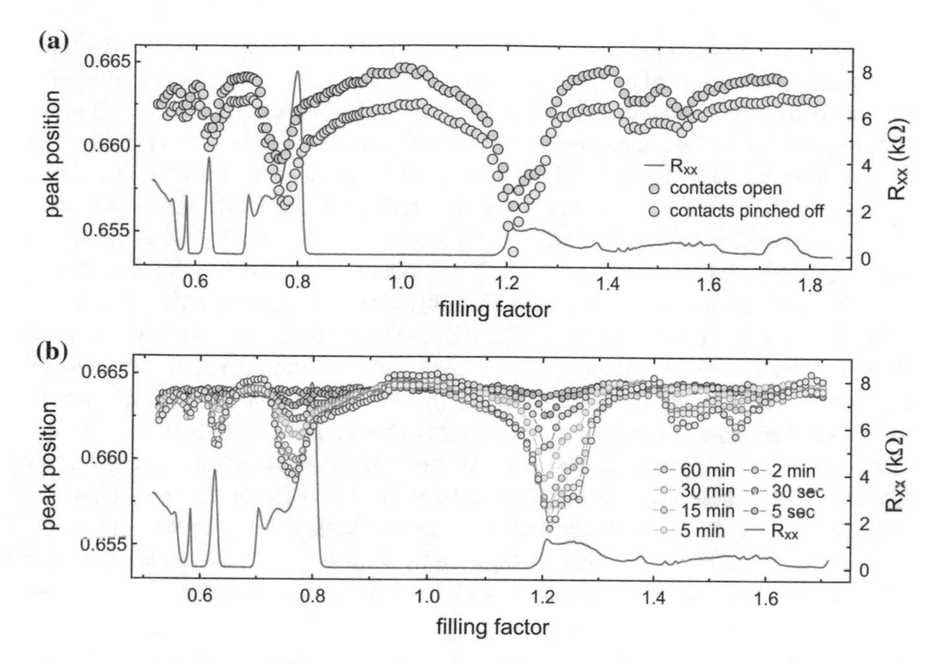


**Fig. 3.13** **a** Filling factor dependence of the nuclear spin polarization with and without pinched-off contacts. Plotted is the peak position (filling factor) of the spin transition at $\nu = 2/3$ ($\propto 1/\mathcal{P}_N$). The measurements were done at base temperature (~20 mK) while applying a magnetic field of 5.55 T. **b** Variation of the relaxation time during which the nuclear spin system approaches thermal equilibrium

Figure 3.13a compares the nuclear spin polarization obtained with and without pinched-off contacts. Also plotted is the longitudinal resistance of the sample. Interestingly, the presence of the topgates modifies the 2DES underneath slightly even if no voltage is applied. As a result, the FQHS around $\nu = 3/2$ appear not fully developed. However, the interior of the sample remains unaffected. It leaps to the eye that the filling factor dependence of $\mathcal{P}_N$ is very similar to Fig. 3.10c albeit measured on a different sample. Small changes presumably arise from the higher magnetic field of 5.55 T used for the measurements in Fig. 3.13a. Of main interest is the observation that both curves exhibit the same variations in $\mathcal{P}_N$ even in the case of pinched-off contacts. Thus, coupling effects via the leads can be ruled out to account for the filling factor dependence of $\mathcal{P}_N$. Interestingly, an overall shift to lower filling factors is observed if the topgates are energized. Apparently, the 2DES is heated slightly by an external current via the leads if not hindered by the topgate. We have studied also the time dependence of the nuclear spin polarization for this sample by varying the waiting time at $\nu_{\text{probe}}$. The outcome is plotted in Fig. 3.13b. It emphasizes that most features in $\mathcal{P}_N$ develop rather slowly, on similar time scales as imposed by the electron-nuclear spin interaction.

Since our findings mostly discard external influences to be responsible for the variations in $\mathcal{P}_N$, the observed behavior is likely to have an intrinsic origin. An intriguing candidate is the ferromagnetic ordering of nuclear spins. Already more than 70 years ago, the nuclear Curie temperature of a three-dimensional metal was calculated to fall into the microkelvin range [46]. However, recently it has been shown that a 2DES with strong electron-electron interactions can exhibit nuclear Curie temperatures on the order of millikelvin [47, 48]. The coupling between the nuclear spins in the 2DES is mainly established via the Rudermann-Kittel-Kasuya-Yosida (RKKY) interaction, i.e. it is mediated indirectly by the electron system. The direct dipolar interaction between nuclei is much weaker [48]. The RKKY interaction requires a compressible electron system to evolve. Hence, only in this case a potential magnetic ordering of the nuclear spins would occur. Consequently, the nuclear spin polarization would be higher in regions of a finite longitudinal resistance. This prediction fits with the main observations in Fig. 3.13. However, further experiments are necessary and in progress to unambiguously identify the origin of the filling-factor-dependent nuclear spin polarization. In any case, our findings provide the possibility to manipulate the nuclear spin polarization simply by changing the filling factor, without the need for strong currents or other external means which raise the temperature.

## References


1. L.W. Engel, S.W. Hwang, T. Sajoto, D.C. Tsui, M. Shayegan, Fractional quantum Hall effect at $\nu = 2/3$ and 3/5 in tilted magnetic fields. Phys. Rev. B **45**, 3418 (1992)
2. I.V. Kukushkin, K. von Klitzing, K. Eberl, Spin polarization of composite fermions: Measurements of the Fermi energy. Phys. Rev. Lett. **82**, 3665 (1999)
3. J.H. Smet, R.A. Deutschmann, W. Wegscheider, G. Abstreiter, K. von Klitzing, Ising ferromagnetism and domain morphology in the fractional quantum Hall regime. Phys. Rev. Lett. **86**, 2412 (2001)
4. V. Piazza, V. Pellegrini, F. Beltram, W. Wegscheider, T. Jungwirth, A.H. MacDonald, First-order phase transitions in a quantum Hall ferromagnet. Nature **402**, 638 (1999)
5. J. Eom, H. Cho, W. Kang, K.L. Campman, A.C. Gossard, M. Bichler, W. Wegscheider, Quantum Hall ferromagnetism in a two-dimensional electron system. Science **289**, 2320 (2000)
6. J. Hayakawa, K. Muraki, G. Yusa, Real-space imaging of fractional quantum Hall liquids. Nat. Nanotechnol. **8**, 31 (2013)
7. S.L. Sondhi, A. Karlhede, S.A. Kivelson, E.H. Rezayi, Skyrmions and the crossover from the integer to fractional quantum Hall effect at small Zeeman energies. Phys. Rev. B **47**, 16419 (1993)
8. S.M. Girvin, Spin and isospin: Exotic order in quantum Hall ferromagnets. Phys. Today **3**, 39 (2000)
9. J.K. Jain, *Composite Fermions* (Cambridge University Press, Cambridge, 2007)
10. Y.Q. Li, J.H. Smet, Nuclear-electron spin interactions in the quantum Hall regime, in *Spin Physics in Semiconductors* (Springer, Berlin, 2008), pp. 347–388
11. B. Verdene, J. Martin, G. Gamez, J. Smet, K. von Klitzing, D. Mahalu, D. Schuh, G. Abstreiter, A. Yacoby, Microscopic manifestation of the spin phase transition at filling factor 2/3. Nat. Phys. **3**, 392 (2007)
12. B.I. Halperin, Theory of the quantized Hall conductance. Helv. Phys. Acta **56**, 75 (1983)
13. J.P. Eisenstein, H.L. Stormer, L.N. Pfeiffer, K.W. West, Evidence for a spin transition in the $\nu = 2/3$ fractional quantum Hall effect. Phys. Rev. B **41**, 7910 (1990)

14. R.G. Clark, S.R. Haynes, A.M. Suckling, J.R. Mallett, P.A. Wright, J.J. Harris, C.T. Foxon, Spin configurations and quasiparticle fractional charge of fractional-quantum-Hall-effect ground states in the N = 0 Landau level. Phys. Rev. Lett. **62**, 1536 (1989)
15. W. Pan, K.W. Baldwin, K.W. West, L.N. Pfeiffer, D.C. Tsui, Spin transition in the $\nu = 8/3$ fractional quantum Hall effect. Phys. Rev. Lett. **108**, 216804 (2012)
16. O. Stern, N. Freytag, A. Fay, W. Dietsche, J.H. Smet, K. von Klitzing, D. Schuh, W. Wegscheider, NMR study of the electron spin polarization in the fractional quantum Hall effect of a single quantum well: Spectroscopic evidence for domain formation. Phys. Rev. B **70**, 075318 (2004)
17. H. Cho, J.B. Young, W. Kang, K.L. Campman, A.C. Gossard, M. Bichler, W. Wegscheider, Hysteresis and spin transitions in the fractional quantum Hall effect. Phys. Rev. Lett. **81**, 2522 (1998)
18. M.J. Snelling, G.P. Flinn, A.S. Plaut, R.T. Harley, A.C. Tropper, R. Eccleston, C.C. Phillips, Magnetic g factor of electrons in GaAs/$Al_xGa_{1-x}As$ quantum wells. Phys. Rev. B **44**, 11345 (1991)
19. S.M. Girvin, A.H. MacDonald, Multicomponent quantum Hall systems: The sum of their parts and more, in *Perspectives in Quantum Hall Effects: Novel Quantum Liquids in Low-dimensional Semiconductor Structures* (John Wiley & Sons, New York, 1996), pp. 161–224
20. T. Jungwirth, A.H. MacDonald, Pseudospin anisotropy classification of quantum Hall ferromagnets. Phys. Rev. B **63**, 035305 (2000)
21. A.H. MacDonald, H.A. Fertig, L. Brey, Skyrmions without sigma models in quantum Hall ferromagnets. Phys. Rev. Lett. **76**, 2153 (1996)
22. T. Jungwirth, S.P. Shukla, L. Smrčka, M. Shayegan, A.H. MacDonald, Magnetic anisotropy in quantum Hall ferromagnets. Phys. Rev. Lett. **81**, 2328 (1998)
23. Y.Q. Li, V. Umansky, K. von Klitzing, J.H. Smet, Current-induced nuclear spin depolarization at Landau level filling factor $\nu = 1/2$. Phys. Rev. B **86**, 115421 (2012)
24. D.R. Lide, *Handbook of Chemistry and Physics* (CRC Press/Taylor & Francis Group, Boca Raton, 2008)
25. M. Dobers, K. von Klitzing, J. Schneider, G. Weimann, K. Ploog, Overhauser-shift of the ESR in the two-dimensional electron gas of GaAs–AlGaAs heterostructures, in *High Magnetic Fields in Semiconductor Physics II* (Springer, Berlin, 1989), pp. 396–400
26. G. Lampel, Nuclear dynamic polarization by optical electronic saturation and optical pumping in semiconductors. Phys. Rev. Lett. **20**, 491 (1968)
27. I.V. Kukushkin, K. von Klitzing, K. Eberl, Enhancement of the skyrmionic excitations due to the suppression of Zeeman energy by optical orientation of nuclear spins. Phys. Rev. B **60**, 2554 (1999)
28. J.H. Smet, R.A. Deutschmann, F. Ertl, W. Wegscheider, G. Abstreiter, K. von Klitzing, Gate-voltage control of spin interactions between electrons and nuclei in a semiconductor. Nature **415**, 281 (2002)
29. K. Hashimoto, K. Muraki, T. Saku, Y. Hirayama, Electrically controlled nuclear spin polarization and relaxation by quantum-Hall states. Phys. Rev. Lett. **88**, 176601 (2002)
30. M. Dobers, K. von Klitzing, J. Schneider, G. Weimann, K. Ploog, Electrical detection of nuclear magnetic resonance in GaAs-$Al_xGa_{1-x}As$ heterostructures. Phys. Rev. Lett. **61**, 1650 (1988)
31. W.D. Knight, Nuclear magnetic resonance shift in metals. Phys. Rev. Lett. **76**, 1529 (1949)
32. S.E. Barrett, G. Dabbagh, L.N. Pfeiffer, K.W. West, R. Tycko, Optically pumped NMR evidence for finite-size skyrmions in GaAs quantum wells near Landau level filling $\nu = 1$. Phys. Rev. Lett. **74**, 5112 (1995)
33. N.N. Kuzma, P. Khandelwal, S.E. Barrett, L.N. Pfeiffer, K.W. West, Ultraslow electron spin dynamics in GaAs quantum wells probed by optically pumped NMR. Science **281**, 686 (1998)
34. A. Wójs, G. Möller, S. Simon, N.R. Cooper, Skyrmions in the Moore–Read state at $\nu = 5/2$. Phys. Rev. Lett. **104**, 086801 (2010)
35. S. Kronmüller, W. Dietsche, J. Weis, K. von Klitzing, W. Wegscheider, M. Bichler, New resistance maxima in the fractional quantum Hall effect regime. Phys. Rev. Lett. **81**, 2526 (1998)

36. S. Kraus, O. Stern, J.G.S. Lok, W. Dietsche, K. von Klitzing, M. Bichler, D. Schuh, W. Wegscheider, From quantum Hall ferromagnetism to huge longitudinal resistance at the 2/3 fractional quantum Hall state. Phys. Rev. Lett. **89**, 266801 (2002)
37. X.Z. Yu, Y. Onose, N. Kanazawa, J.H. Park, J.H. Han, Y. Matsui, N. Nagaosa, Y. Tokura, Real-space observation of a two-dimensional skyrmion crystal. Nature **465**, 901 (2010)
38. R. Tycko, S.E. Barrett, G. Dabbagh, L.N. Pfeiffer, K.W. West, Electronic states in gallium arsenide quantum wells probed by optically pumped NMR. Science **268**, 1460 (1995)
39. L. Brey, H.A. Fertig, R. Côté, A.H. MacDonald, Skyrme crystal in a two-dimensional electron gas. Phys. Rev. Lett. **75**, 2562 (1995)
40. R. Côté, A.H. MacDonald, L. Brey, H.A. Fertig, S.M. Girvin, H.T.C. Stoof, Collective excitations, NMR, and phase transitions in Skyrme crystals. Phys. Rev. Lett. **78**, 4825 (1997)
41. V. Bayot, E. Grivei, S. Melinte, M.B. Santos, M. Shayegan, Giant low temperature heat capacity of GaAs quantum wells near Landau level filling $\nu = 1$. Phys. Rev. Lett. **76**, 4584 (1996)
42. E.H. Aifer, B.B. Goldberg, D.A. Broido, Evidence of skyrmion excitations about $\nu = 1$ in n-modulation-doped single quantum wells by interband optical transmission. Phys. Rev. Lett. **76**, 680 (1996)
43. L.A. Tracy, J.P. Eisenstein, L.N. Pfeiffer, K.W. West, Resistively detected NMR in a two-dimensional electron system near $\nu = 1$: Clues to the origin of the dispersive lineshape. Phys. Rev. B **73**, 121306 (2006)
44. G. Gervais, H.L. Stormer, D.C. Tsui, P.L. Kuhns, W.G. Moulton, A.P. Reyes, L.N. Pfeiffer, K.W. Baldwin, K.W. West, Evidence for skyrmion crystallization from NMR relaxation experiments. Phys. Rev. Lett. **94**, 196803 (2005)
45. J.H. Smet, R.A. Deutschmann, F. Ertl, W. Wegscheider, G. Abstreiter, K. von Klitzing, Anomalous-filling-factor-dependent nuclear-spin polarization in a 2D electron system. Phys. Rev. Lett. **92**, 086802 (2004)
46. H. Fröhlich, F.R.N. Nabarro, Orientation of nuclear spins in metals. Proc. R. Soc. Lond. A **175**, 382 (1940)
47. P. Simon, D. Loss, Nuclear spin ferromagnetic phase transition in an interacting two dimensional electron gas. Phys. Rev. Lett. **98**, 156401 (2007)
48. P. Simon, B. Braunecker, D. Loss, Magnetic ordering of nuclear spins in an interacting two-dimensional electron gas. Phys. Rev. B **77**, 045108 (2008)

# Chapter 4
# The Spin Polarization of the $^5/_2$ State

In the previous chapter, we exploited the hyperfine coupling of electrons and nuclei to investigate different electronic spin excitations. In this chapter, we rely on a different aspect of the hyperfine interaction, namely the ability to probe the electron spin polarization by a nuclear magnetic resonance (NMR) of the host crystal. Using this technique, we have measured the spin polarization of the FQHS at filling factor $\nu = {}^5/_2$. As discussed in Sect. 2.6.2, it is a crucial quantity to unravel the enigmatic nature of the $\nu = {}^5/_2$ state.

## 4.1 Resistively Detected NMR

For performing NMR experiments, a coil was wound around the sample as depicted in Fig. 4.1. It allows applying an oscillating magnetic field with frequency $f$ in the plane of the 2DES. If the corresponding photon energy $hf$ is tuned such that it matches the Zeeman splitting of the nuclei, transitions between nuclear energy levels can be triggered by photon absorption (Fig. 4.2a). For short interaction times, this can be used to coherently control the nuclear spin polarization. If the radio frequency (RF) radiation is applied over a long time scale, in particular longer than the dephasing time, the coherence is lost, and the mean nuclear spin polarization $\mathcal{P}_N$ vanishes. This is referred to as the continuous-wave mode. The decrease of nuclear spin polarization under resonant excitation affects the energetic situation of the electron system. As mentioned earlier (Sect. 3.1), a spin-polarized nuclear spin system changes the electronic Zeeman energy via the hyperfine interaction. This contribution to the electronic Zeeman energy is often expressed in terms of a fictitious magnetic field, the Overhauser field $B_N$, acting on the electron system as $E_z = g_e^* \mu_B (B + B_N)$. The Overhauser field is oriented opposite to the external magnetic field. Thus, a polarized nuclear spin system reduces the Zeeman splitting (Fig. 4.2b). As a result, the hyperfine coupling lays the foundation to conveniently identify the nuclear resonance condition by a change in the longitudinal resistance of the sample if the filling factor is chosen properly—a method named resistively detected NMR. In general, a strong



B. Frieß, *Spin and Charge Ordering in the Quantum Hall Regime*,
Springer Theses, DOI 10.1007/978-3-319-33536-0_4

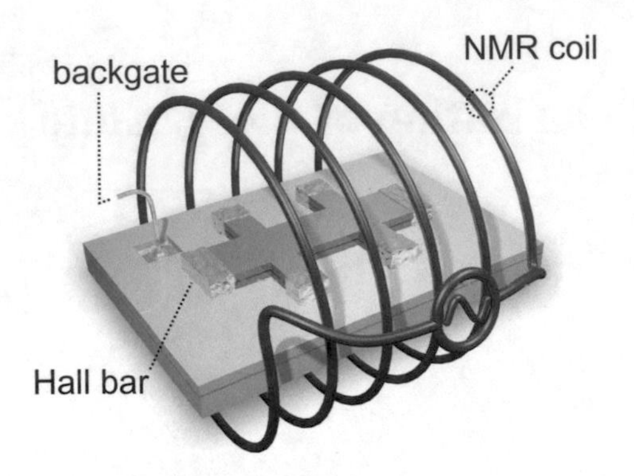


**Fig. 4.1** Experimental setup for the NMR experiments

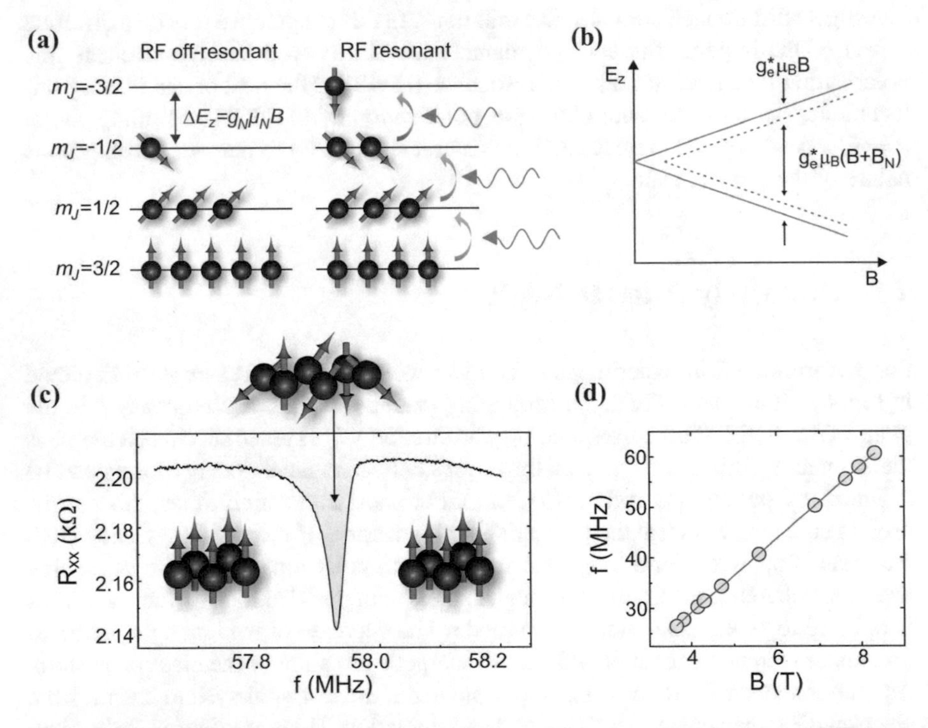


**Fig. 4.2** **a** Resonant RF radiation excites nuclear spins to higher energy levels and thereby reduces the nuclear spin polarization. Shown is the case for nuclei with spin $\mathcal{J} = 3/2$. **b** The fictitious Overhauser field $B_N$ arises from a finite nuclear spin polarization and causes a reduction of the electronic Zeeman energy. **c** Resistively detected NMR spectrum of $^{75}$As nuclei. The radio frequency was swept from high to low values while recording the longitudinal resistance. **d** Nuclear resonance frequency measured at different magnetic fields. A linear fit yields the gyromagnetic ratio of the $^{75}$As nuclei

response in $R_{xx}$ is expected for filling factors at the flank of a quantum Hall state whose energy gap depends predominantly on the Zeeman energy, e.g. at the flank of $\nu = 1$. Figure 4.2c gives an example of a nuclear magnetic resonance detected via the longitudinal resistance of the sample. It shows a measurement of $R_{xx}$ as the radio frequency was swept across the nuclear resonance. At high frequencies, when the radio frequency is off-resonant to the nuclear Zeeman splitting, $\mathcal{P}_N$ has a finite value corresponding to the relative occupation of the nuclear energy levels. If the RF energy is tuned into resonance, nuclei get excited to higher energy levels, and as a consequence the average spin polarization decreases. In the present case, this causes a drastic change in $R_{xx}$ to lower values. Once the RF radiation is off-resonant again, $\mathcal{P}_N$ recovers thermal equilibrium at a time scale set by $T_1$. At the same time, $R_{xx}$ returns to its equilibrium value. The resonance spectrum in Fig. 4.2c was taken for $^{75}$As nuclei. Throughout this thesis, we exclusively studied resonances of $^{75}$As isotopes as they yield the strongest signal owing to their high abundance. In order to unambiguously identify the feature in $R_{xx}$ as a nuclear magnetic resonance, we have analyzed the magnetic field dependence of the resonance frequency as shown in Fig. 4.2d. The resonance shifts linearly with a slope of $\gamma/2\pi = 7.29\,\text{MHz/T}$ when changing the magnetic field. This value is in good agreement with the reduced gyromagnetic ratio of $^{75}$As nuclei (see Table 3.1).

Figure 4.2c nicely demonstrates that a nuclear magnetic resonance can be detected via the longitudinal resistance. Yet, the question remains how this can be utilized to measure the spin polarization of the electron system. The key ingredient to answer this question is again the hyperfine interaction. In the same way as a polarized nuclear spin system changes the Zeeman energy of the electrons, also a spin-polarized electron system influences the Zeeman splitting of the nuclei. Thus, a non-zero value of the electron spin polarization manifests itself as a linear shift of the NMR frequency, introduced in Sect. 3.1.3 as the Knight shift $K_s$ [1]. The Knight shift depends on both the spin polarization and the electron density—overall, the electron spin density. Over the last years, this technique has been established as a reliable method to probe the electron spin polarization [2–5]. The downside of this method is that it is only directly applicable at filling factors where $R_{xx}$ depends on the Zeeman energy. In many cases this requirement is not fulfilled, especially not for quantum Hall states where $R_{xx} \approx 0$. We have solved this issue by taking advantage of the backgate present in our sample.

The extended measurement procedure is shown in Fig. 4.3. It divides the measurement process into two parts, separating the manipulation of the nuclei from the detection of the nuclear spin polarization. In the first step, a radio frequency close to the nuclear resonance frequency is applied at $\nu_{\text{probe}}$. It is the electron spin polarization of this filling factor which determines the Knight shift. In the second step, the RF-induced change of $\mathcal{P}_N$ is detected by measuring $R_{xx}$ at a different filling factor $\nu_{\text{detect}}$. These two steps are repeated multiple times while for each cycle the applied frequency is lowered slightly to scan across the whole resonance. With this method, resonances can be acquired in principle at every filling factor $\nu_{\text{probe}}$ within the tuning range of the backgate. The magnetic field is kept constant throughout the entire sequence. An important prerequisite for this measurement technique is the

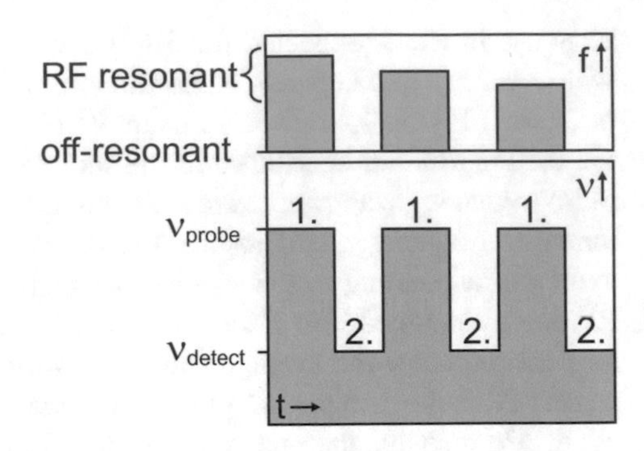


**Fig. 4.3** Measurement sequence used to acquire NMR spectra at different filling factors $\nu_{\text{probe}}$. The technique relies on abrupt changes of the filling factor to separate the RF-induced nuclear spin manipulation from its detection (see text for details)

preservation of the nuclear spin polarization during read out. This is ensured firstly by a swift control of the filling factor on time scales much faster than the nuclear spin relaxation time. Secondly, the RF radiation is set off-resonant during read-out to avoid a further manipulation of the nuclei. The same effect can be achieved by turning off the RF power during read-out. However, this would lead to a temperature change with sometimes unforeseeable side effects. The described measurement technique for the determination of the electron spin polarization requires the nuclear spin system to be at least partially polarized. This can be achieved either by dynamic nuclear polarization [6] or, as in our case, by sufficiently low temperatures (see Fig. 3.3b). The need for ultra-low temperatures is inherently counteracted by the unavoidable heating due to the applied RF radiation. On the one hand, higher RF powers improve the NMR signal. On the other hand, the concomitant temperature change reduces $\mathcal{P}_N$. The heating effects are in particular problematic when probing fragile quantum Hall states, such as the $\nu = 5/2$ state. Here, low RF powers are necessary to impair the quality of the FQHS as little as possible, rendering signal acquisition a challenge. In view of the conflicting priorities of low temperatures and good signal-to-noise ratio, we maximized the NMR signal by averaging over multiple measurement cycles.

## 4.2 Results and Discussion

In order to determine the spin polarization at $\nu = 5/2$, first the Knight shift needs to be calibrated. This is done by recording NMR spectra at quantum Hall states with a known spin polarization. For this purpose, we have measured resonance spectra at the filling factors $\nu = 2$, 1 and $5/3$ (Fig. 4.4a–c). In the case of $\nu = 2$, both spin branches of the first Landau level are completely filled. Hence, the electron system at this first reference point is unpolarized. For $\nu = 1$, in contrast, only the lower spin branch of the first Landau level is occupied, and the electron system is considered to be maximally polarized. As a third reference point, $\nu = 5/3$ is chosen. It is the particle-hole-conjugate state of $\nu = 1/3$ with respect to $\nu = 2$ and is therefore spin polarized. For the following considerations, it is important to bear in mind that

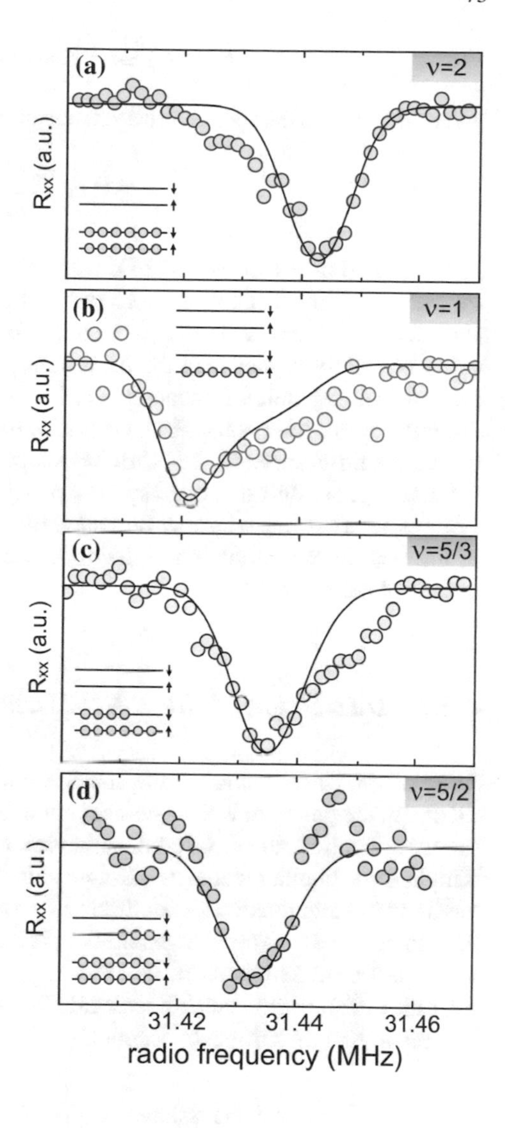


**Fig. 4.4** Resistively detected NMR spectra of $^{75}$As nuclei measured according to Fig. 4.3 at filling factors $\nu = 2$ (**a**), $\nu = 1$ (**b**), $\nu = 5/3$ (**c**) and $\nu = 5/2$ (**d**). The magnetic field was kept constant at 4.3 T. The insets show the nominal Landau level occupation. The solid lines represent fits to theory

all measurements in Fig. 4.4 were obtained at a constant magnetic field of 4.3 T. Thus, the shifts to lower frequencies appearing at $\nu = 1$ and $\nu = 5/3$ stem solely from the non-zero electron spin polarization and represent the Knight shift. The frequency difference $f_{\nu=2} - f_{\nu=1} = 22$ kHz delineates the maximum possible Knight shift $K_{s,max}$ for the present (fixed) magnetic field. Its exact value is magnetic field dependent since the Landau level degeneracy and therefore also the maximum spin density at full polarization changes for different $B$-fields. For a known electron spin polarization and at a constant magnetic field, the corresponding Knight shift can in principle be calculated according to

$$K_s(\mathcal{P}^*) = (f_{\nu=2} - f_{\nu=1}) \cdot \mathcal{P}^* = K_{s,max} \cdot \mathcal{P}^*, \tag{4.1}$$

where $\mathcal{P}^*$ denotes the spin polarization normalized by the Landau level degeneracy $n_L$

$$\mathcal{P}^* = \frac{n_\uparrow - n_\downarrow}{n_L}. \tag{4.2}$$

In turn, relation 4.1 allows us to directly determine the electron spin polarization by measuring the Knight shift. However, applying this calibration to the Knight shift $K_s = 8.4\,\text{kHz}$ at $\nu = 5/3$ yields a spin polarization of $\mathcal{P}^* = 39\,\%$. This value is slightly higher than the maximum polarization possible at $\nu = 5/3$ in the case of well-separated Landau levels, i.e. $\mathcal{P}^* = 1/3$. We will come back to this discrepancy at a later point. First, we briefly touch upon a different topic. It addresses the salient differences of the NMR lineshapes in Fig. 4.4. The resonance at $\nu = 1$, for instance, exhibits a strong asymmetry, whereas at $\nu = 2$ the resonance is rather symmetric. This behavior can be understood as a consequence of the finite width of the electron wavefunction in the direction perpendicular to the quantum well as described below [4, 7].

### 4.2.1 *Discussion of the NMR Lineshape*

The electron wavefunction of the 2DES is spatially extended in the direction perpendicular to the quantum well as pointed out in Sect. 2.1. Owing to the spatial spread of the wavefunction, the electron density varies along the $z$-axis and so does the local Knight shift. In other words, the Knight shift of a single nucleus depends on its local environment and thus varies for different positions in the quantum well. The overall NMR lineshape measured in experiment is therefore composed of single resonances with a different Knight shift. Its exact shape $\mathcal{I}(f)$ can be calculated by integrating over the different resonance lines while taking the local electron densities via their respective Knight shifts into account:

$$\mathcal{I}(f) = \int \text{Gaussian}(f - (f_0 - \xi\,\mathcal{P}^*\,|\psi(z)|^2))\,|\psi_{\text{detect}}(z)|^2\,dz, \tag{4.3}$$

where $f_0$ denotes the unshifted resonance frequency and $\psi(z)$ represents the electron wavefunction in the $z$-direction. It was assumed that a single nucleus has a Gaussian lineshape. Since our measurement technique relies on different filling factors for detection and RF excitation, the distinct electron distribution present at $\nu_{\text{detect}}$ needs to be taken into account as well. This is done by weighting the single Gaussian functions with $|\psi_{\text{detect}}(z)|^2$ while assuming that regions of higher electron density contribute stronger to the resistive detection. The calibration constant $\xi$ defines how much the nuclear resonance frequency is shifted by a given electron spin polarization. It is magnetic field dependent and equals $K_{s,max}$ in the limit of narrow wavefunctions.

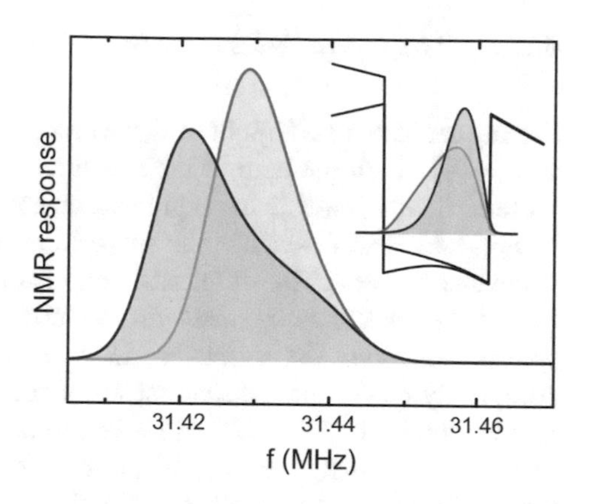


**Fig. 4.5** Simulated NMR response at $\nu = 1$ (*red*) and $\nu = 5/2$ (*blue*) when assuming equal spin polarization and electron density. The inset shows the shape of the wavefunction (probability density) for each of the two filling factors

For realistic experimental conditions, the definition of the Knight shift given in Eq. 4.1 is problematic since already the varying shape of the wavefunction at different filling factors can shift the overall resonance frequency, even in the case of equal spin densities. This behavior is due to the fact that the filling factor is tuned electrostatically. The shape of the quantum well and consequently also the wavefunction changes for different backgate voltages. This effect is illustrated in Fig. 4.5. It shows the wavefunctions at two different filling factors and the resulting resonances when assuming equal spin polarization and electron density. Interestingly, a symmetric density distribution in the quantum well causes an asymmetric resonance and vice versa.

Based on Eq. 4.3, we calculated the resonance spectra for the experimental data in Fig. 4.4. The shape of the wavefunction at the respective filling factors was determined by solving the Schrödinger and Poisson equations iteratively using the software *nextnano++* [8]. The resonance at $\nu = 2$ is fit simply by a Gaussian lineshape: In the case of an unpolarized electron system the Knight shift is zero, and the resonance remains symmetric irrespective of the density distribution. Filling factor $\nu = 5/3$ was used as a second reference point. The electron spin polarization was fixed to $\mathcal{P}^* = 1/3$, and $\xi$ was varied to yield best fit. The shoulder which appears on the high frequency side is not reproduced by the fit. Possibly, some of the structural parameters, such as the quantum well width, differ from their nominal value. With the value of $\xi$ determined for $\nu = 5/3$, we now turn to the resonance at $\nu = 1$. Here, the electron spin polarization was varied as the only parameter to fit the resonance. The result is $\mathcal{P}^*(\nu = 1) = 0.73 \pm 0.17\,\%$, which is considerably smaller than the expected full polarization ($\mathcal{P}^* = 1$). This discrepancy, already deduced earlier from the ratio of the Knight shifts, cannot be explained by the different shape of the wavefunctions. Nevertheless, we deem the calibration at $\nu = 5/3$ reliable. A spin polarization $\mathcal{P}^*(\nu = 1) < 1$ is not surprising in view of the skyrmions present in this region (see Sect. 3.2.2) and has been observed before [4, 9]. In turn, using $\mathcal{P}^*(\nu = 1) = 1$ as a calibration would imply $\mathcal{P}^*(\nu = 5/3) > 1/3$, i.e. the spin polarization of the $5/3$ state would be higher than its maximum value.

### 4.2.2 The Spin Polarization at $\nu = {}^5\!/\!_2$

Having calibrated the Knight shift, we proceed with determining the spin polarization at $\nu = {}^5\!/\!_2$. With the help of the measurement sequence described in the previous section, it was possible to acquire resonance spectra also at $\nu = {}^5\!/\!_2$ as depicted in Fig. 4.4d. The resonance lineshape at $\nu = {}^5\!/\!_2$ is symmetric in contrast to the resonance at $\nu = 1$. This difference stems from the altered shape of the wavefunction. Fitting the nuclear resonance with the model presented above reveals the electron spin polarization. The simulated NMR response in Fig. 4.4d reflects the case of a completely spin-polarized electron system in the half-filled spin branch of the second Landau level, i.e. $\mathcal{P}^* = 0.5$. However, to be able to attribute this spin polarization to the $\nu = {}^5\!/\!_2$ FHQS, it is crucial to consider the effect of RF-induced heating on the 2DES. In Fig. 4.6a the longitudinal resistance between $\nu = 2$ and $\nu = 2.6$ is plotted, measured under the influence of four different RF powers as well as at base temperature. Of course, the heating caused by high RF powers degrades the quality of the $^5\!/\!_2$ state. To determine the exact value of the RF-induced temperature change, we recorded $R_{xx}$ in the same filling factor range at different temperatures (Fig. 4.6b). From the temperature dependence of the resistance value at $\nu = {}^5\!/\!_2$, the energy gap $\Delta_{5/2}$ of the $^5\!/\!_2$ state can be extracted. This is done best with the aid of an Arrhenius plot as shown in Fig. 4.6c while assuming $R(T) \propto \exp(-\frac{\Delta}{2k_BT})$. A linear fit in the temperature-activated regime yields an energy gap $\Delta_{5/2} = 168\,\text{mK}$. Knowing the temperature dependence of $R_{xx}$, the resistance values in Fig. 4.6a can be converted to the corresponding electron temperature. The result is depicted in Fig. 4.6d.

The resonance at $\nu = {}^5\!/\!_2$ shown in Fig. 4.4d was measured at a RF power of $-20\,\text{dBm}$. Thanks to the excellent sensitivity of the resistive detection method, it was possible to reduce the RF power further and acquire resonance spectra down to $-50\,\text{dBm}$. The extracted values of $\mathcal{P}^*(\nu = {}^5\!/\!_2)$ are summarized in Fig. 4.7. The error bars comprise the uncertainty in determining the resonance frequencies. At all accessible temperatures, $\mathcal{P}^*(\nu = {}^5\!/\!_2)$ is consistent with a completely polarized electron system.

At high temperatures, when the $^5\!/\!_2$ state has not yet developed, a fully polarized electron system fits with the expectations of the non-interacting electron picture. All electrons within a Landau level spin branch have the same spin orientation provided that the Zeeman splitting is strong enough. However, if the CF picture of the first Landau level (Sect. 2.4.2) is extended to the second Landau level, it predicts a Fermi sea of composite fermions at half filling. In this case, the spin polarization is determined by the ratio of the Zeeman and CF Fermi energy as described in Sect. 3.1.1. Hence, a spin-polarized electron system requires

$$E_F = \frac{\hbar^2 k_{CF}^2}{2m_{CF}^*} = \frac{\hbar^2 2\pi n_{CF}}{2m_{CF}^*} > g_e \mu_B B, \tag{4.4}$$

where $n_{CF}$ is the density of composite fermions, i.e. $n_{CF} = n_e/5$ in the case of $\nu = {}^5\!/\!_2$. For the present experimental conditions, this implies a CF mass $m_{CF}^* > 1.14\,m_e$,

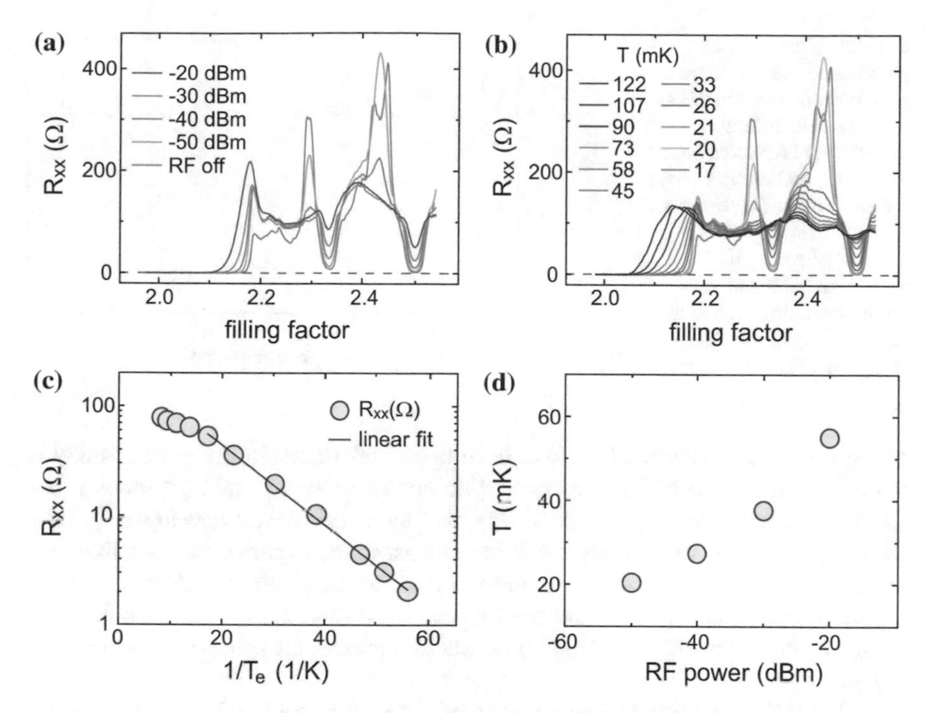


**Fig. 4.6** Estimation of the RF-induced heating. **a** Longitudinal resistance measured at 4.3 T for different RF powers. **b** $R_{xx}$ in the same filling factor range for different temperatures. **c** Arrhenius plot of the resistance values at $\nu = {}^5/_2$ with linear fit. **d** Conversion of RF power to electron temperature

which is close to the value $m^*_{CF} = 1.13\, m_e$ determined by Kukushkin et al. at filling factor $\nu = {}^3/_2$ [10].[1] While the validity of the CF picture in higher Landau levels is questionable, indications of a CF Fermi sea at $\nu = {}^5/_2$ have been found by surface acoustic wave experiments [11].

At lower temperatures, when $R_{xx}(\nu = {}^5/_2)$ approaches zero, Fig. 4.7 unveils the sought for spin polarization of the ${}^5/_2$ state. The discovery of a fully polarized state confirms the prediction of the Moore-Read theory. Together with the observed $e/4$ fractional charge [12–14] these results strongly favor the (anti-)Pfaffian wavefunction as a valid description of the ${}^5/_2$ state [15–17]. This would further imply the non-abelian nature of the ${}^5/_2$ quasiparticles [18–20]. Our results exclude the unpolarized 331 state [21], which was considered the most likely abelian contender in explaining the existence of the ${}^5/_2$ state. For a discussion of the main candidate wavefunctions and a review on the experiments performed so far, we refer to Sect. 2.6.2.

In the course of acquiring these results, two similar pieces of work were published by other groups [4, 5]. The authors also inferred a spin-polarized ${}^5/_2$ state in consistence with our findings. What distinguishes the present work is the quality

[1] It was assumed that the CF mass scales with the magnetic field according to $m^*_{CF} \propto \sqrt{B} m_e$ [10].

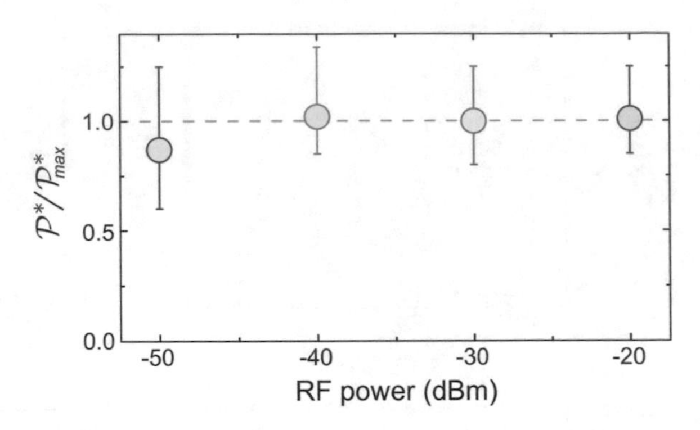


**Fig. 4.7** Electron spin polarization measured by resistively detected NMR at $\nu = 5/2$ with different intensities of RF radiation. The transport behavior under these conditions is shown in Fig. 4.6a. The values of $\mathcal{P}^*$ are normalized by the maximum polarization possible within a spin-split Landau level at $\nu = 5/2$ ($P^*_{max} = 0.5$)

of the 5/2 state. Considerable efforts have been undertaken in the present study to optimize the cooling of the electrons while keeping the external heat input at bay (details can be found in appendix B). Figure 4.6a reveals a well-developed 5/2 state with $R_{xx}$ becoming vanishingly small in contrast to the experimental conditions in references [4, 5]. This is an important point since the electron system at $\nu = 5/2$ is spin polarized also in the absence of a quantized state. It is therefore difficult to judge at which temperature $\mathcal{P}(\nu = 5/2)$ starts to represent the spin polarization of the FQHS.

In Fig. 4.7 a tendency to lower values of $\mathcal{P}^*$ is observable when decreasing the temperature. With the present accuracy, however, it is not possible to state whether this is a genuine trend or not. In principle, our results would also support a high partial spin polarization. For this depth of analysis, additional experiments will be necessary to either increase the measurement accuracy or improve the quality of the $\nu = 5/2$ state further.

## References

1. W.D. Knight, Nuclear magnetic resonance shift in metals. Phys. Rev. Lett. **76**, 1529 (1949)
2. W. Desrat, D.K. Maude, M. Potemski, J.C. Portal, Z.R. Wasilewski, G. Hill, Resistively detected nuclear magnetic resonance in the quantum Hall regime: Possible evidence for a Skyrme crystal. Phys. Rev. Lett. **88**, 256807 (2002)
3. O. Stern, N. Freytag, A. Fay, W. Dietsche, J.H. Smet, K. von Klitzing, D. Schuh, W. Wegscheider, NMR study of the electron spin polarization in the fractional quantum Hall effect of a single quantum well: Spectroscopic evidence for domain formation. Phys. Rev. B **70**, 075318 (2004)
4. L. Tiemann, G. Gamez, N. Kumada, K. Muraki, Unraveling the spin polarization of the $\nu = 5/2$ fractional quantum Hall state. Science **335**, 828 (2012)
5. M. Stern, B.A. Piot, Y. Vardi, V. Umansky, P. Plochocka, D.K. Maude, I. Bar-Joseph, NMR probing of the spin polarization of the $\nu = 5/2$ quantum Hall state. Phys. Rev. Lett. **108**, 066810 (2012)
6. M. Dobers, K. von Klitzing, J. Schneider, G. Weimann, K. Ploog, Electrical detection of nuclear magnetic resonance in GaAs-$Al_xGa_{1-x}$ As heterostructures. Phys. Rev. Lett. **61**, 1650 (1988)

7. N.N. Kuzma, P. Khandelwal, S.E. Barrett, L.N. Pfeiffer, K.W. West, Ultraslow electron spin dynamics in GaAs quantum wells probed by optically pumped NMR. Science **281**, 686 (1998)
8. S. Birner, T. Zibold, T. Andlauer, T. Kubis, M. Sabathil, A. Trellakis, P. Vogl, Nextnano: General purpose 3-D simulations. IEEE Trans. Elect. Devices **54**, 2137 (2007)
9. E.H. Aifer, B.B. Goldberg, D.A. Broido, Evidence of skyrmion excitations about $\nu = 1$ in n-modulation-doped single quantum wells by interband optical transmission. Phys. Rev. Lett. **76**, 680 (1996)
10. I.V. Kukushkin, K. von Klitzing, K. Eberl, Spin polarization of composite fermions: Measurements of the Fermi energy. Phys. Rev. Lett. **82**, 3665 (1999)
11. R.L. Willett, K.W. West, L.N. Pfeiffer, Experimental demonstration of Fermi surface effects at filling factor 5/2. Phys. Rev. Lett. **88**, 066801 (2002)
12. I.P. Radu, J.B. Miller, C.M. Marcus, M.A. Kastner, L.N. Pfeiffer, K.W. West, Quasi-particle properties from tunneling in the $\nu = 5/2$ fractional quantum Hall state. Science **320**, 899 (2008)
13. M. Dolev, M. Heiblum, V. Umansky, A. Stern, D. Mahalu, Observation of a quarter of an electron charge at the $\nu = 5/2$ quantum Hall state. Nature **452**, 829 (2008)
14. V. Venkatachalam, A. Yacoby, L. Pfeiffer, K. West, Local charge of the $\nu = 5/2$ fractional quantum Hall state. Nature **469**, 185 (2011)
15. R.H. Morf, Transition from quantum Hall to compressible states in the second Landau level: New light on the $\nu = 5/2$ enigma. Phys. Rev. Lett. **80**, 1505 (1998)
16. A.E. Feiguin, E. Rezayi, K. Yang, C. Nayak, S. Das Sarma, Spin polarization of the $\nu = 5/2$ quantum Hall state. Phys. Rev. B **79**, 115322 (2009)
17. J.K. Jain, The 5/2 enigma in a spin? Physics **3**, 71 (2010)
18. G. Moore, N. Read, Nonabelions in the fractional quantum Hall effect. Nucl. Phys. B **360**, 362 (1991)
19. M. Greiter, X.G. Wen, F. Wilczek, Paired Hall states. Nucl. Phys. B **374**, 567 (1992)
20. A. Stern, Non-abelian states of matter. Nature **464**, 187 (2010)
21. B.I. Halperin, Theory of the quantized Hall conductance. Helv. Phys. Acta **56**, 75 (1983)

# Chapter 5
# Probing the Microscopic Structure of the Stripe Phase at $\nu = 5/2$

In the previous chapter, we probed the electron spin polarization by means of resistively detected NMR. It was shown that the resonance lineshape not only depends on the spin polarization but also on the spatial distribution of the electrons inside of the quantum well. This technique can therefore be used more generally to probe the distribution of the electron density in all three spatial dimensions. Due to this local sensitivity, NMR is ideally suited to study the spatial ordering of electrons which is believed to occur in the density-modulated phases introduced in Sect. 2.5. This constitutes the central topic of the present chapter. More to the point, it deals with the study of the stripe phase emerging at filling factor $\nu = 5/2$ when rotating the sample with respect to the external magnetic field. The first section gives an overview of the experimental indications for such a stripe phase. In the second part, we employ the NMR technique to study the spatial density distribution of this phase. The results are discussed in the last section, underpinned by a theoretical model of the stripe formation. From this model, the stripe period as well as the modulation strength are deduced.

## 5.1 Tilt-Induced Phase Transition at $\nu = 5/2$

The appearance of density-modulated phases in the quantum Hall regime was introduced in Sect. 2.5. For such phases, the homogeneous electron system in the topmost Landau level is assumed to break up into regions of alternating electron density. Their existence is inferred from characteristic features in the transport behavior. In the case of the bubble phase, a reappearance of the IQHE indicates a pinning of the excess electrons in the partially occupied Landau level. More precisely, the electrons are believed to cluster in bubbles with an integer filling factor, which arrange in a triangular lattice. This distinguishes the bubble phase from the stripe phase. In case of the stripe phase, electrons are supposed to order in stripe-like patterns of alternating filling factor. The appearance of a strong transport anisotropy at half Landau level fillings is construed as an indication for such a stripe phase.



B. Frieß, *Spin and Charge Ordering in the Quantum Hall Regime*,
Springer Theses, DOI 10.1007/978-3-319-33536-0_5

These density-modulated phases compete at partial fillings with FQHS for stability. In the first Landau level FQHS are ubiquitous, whereas in higher Landau levels density-modulated phases prevail over the FQHE. This competition stems from subtle changes in the direct Coulomb and exchange interaction owing to the altered shape and extent of the electron wavefunction in the respective Landau levels. As already addressed in Sect. 2.6, this effect looms largest in the second Landau level, where FQHS are intertwined with reentrant integer quantum Hall states (see Fig. 2.19). Here, transitions between FQHS and density-modulated phases are induced by minute changes of the electron density or the magnetic field. The close proximity of FQHS and density-modulated phases further manifests itself in Fig. 5.1. It shows the longitudinal resistance along two perpendicular current directions under the influence of an additional magnetic field component in the plane of the 2DES. The measurement was performed at about 20 mK in a dilution refrigerator. To allow for a proper comparison between both current directions, the sample was patterned in a square geometry (400 µm wide). The in-plane field component was created by tilting the sample with respect to the external magnetic field. The filling factor is set by the perpendicular field component $B_{\perp} = \cos(\varphi) B_{\text{tot}}$, where $B_{\text{tot}}$ denotes the total magnetic field and $\varphi$ the angle spanned by $B_{\text{tot}}$ and $B_{\perp}$. Figure 5.1 reveals two salient observations. Firstly, upon tilting of the sample the $5/2$ state gradually disappears, while the states at $\nu = 7/3$ and $\nu = 8/3$ get strengthened by the in-plane magnetic field. Secondly, the weakening of the $5/2$ state is accompanied by an emergent strong transport anisotropy. Its hard axis is oriented along the in-plane magnetic field component $B_{\parallel}$. These observations are consistent with previous publications [1–4].

The disappearance of the $5/2$ state had first been taken as evidence for an unpolarized electron system since an in-plane magnetic field selectively enhances the Zeeman energy [1]. In view of the strong transport anisotropy, this interpretation has later been revised in favor of a stripe phase formation. According to Hartree–Fock calculations, the anisotropic phase may be understood as a unidirectional charge density wave (CDW) [5–7]. In this case, the local filling factor in the topmost Landau level alternates between two neighboring integer numbers in stripes with a strictly one-dimensional periodicity. This behavior is sketched in Fig. 5.2a. Alternative models take fluctuations along the stripes into account and approach the stripe phase in close analogy to the behavior of liquid crystals [8, 9]. Fradkin and Kivelson proposed distinct electron liquid crystal phases which are categorized according to their symmetry and strength of shape fluctuations [8]: the smectic, nematic and stripe crystal phase. Details on these phases can be found in references [8, 9]. Based on this literature, we briefly introduce the different electron liquid phases in the following:

- Hartree–Fock calculations indicate that the CDW at zero temperature is generally unstable to the formation of modulations along the stripes [10]. In the case of the **stripe crystal phase**, these modulations order in an anti-phase manner (Fig. 5.2b). The resultant phase is equivalent to an anisotropic Wigner crystal.
- The **smectic phase** has modulations along the stripes similar to the stripe crystal, albeit with no long range anti-phase order due to dynamic phase slips (Fig. 5.2c).

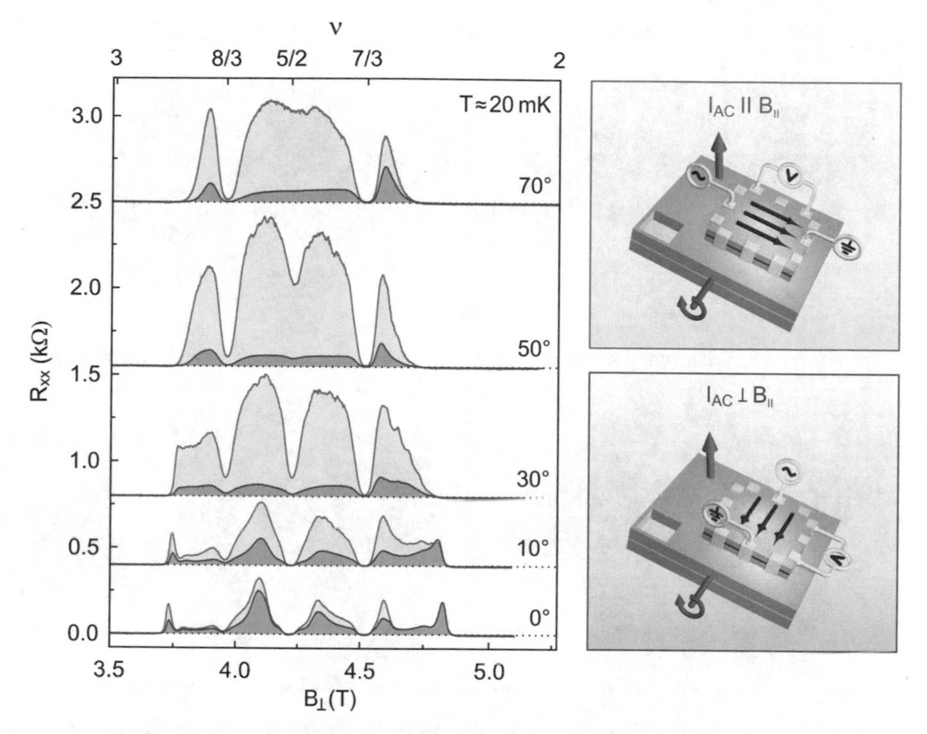


**Fig. 5.1** Influence of an in-plane magnetic field $B_{\|}$ on the longitudinal resistance (measured at $T \approx 20\,\text{mK}$). The sample was tilted by an angle $\varphi$ with respect to the external magnetic field. $\varphi = 0°$ corresponds to $B_{\|} = 0$. The transport was probed along two perpendicular current directions: $I_{\text{AC}} \parallel B_{\|}$ (*blue*) and $I_{\text{AC}} \perp B_{\|}$ (*red*). The direction of current flow is indicated by *red arrows*

In the thermodynamic limit, the smectic phase is equivalent to a CDW with no modulations. An important prerequisite for smectic order is the continuity of the stripes. The amplitude of the shape fluctuations is small compared to the stripe period such that neighboring stripes do not overlap. This requirement distinguishes the smectic phase from the nematic phase.

- In the case of the **nematic phase**, the fluctuations along the stripes are strong enough that stripes may break apart and dislocations can occur (Fig. 5.2d). On the other hand, the fluctuations must not be too strong so that the orientational order of the stripes persists. Otherwise, the anisotropic character of the nematic phase would be lost.

Beyond the mere transport behavior, experiments on the local nature of the anisotropic phases in the quantum Hall regime are scarce. The temperature dependence of the transport anisotropy at $\nu = 9/2$ has been analyzed and was found to be consistent with the predictions of a nematic phase [11, 12]. Apart from that, resonances in the microwave absorption have been observed and interpreted as pinning modes of the stripe crystal [13, 14]. Yet, with these techniques microscopic

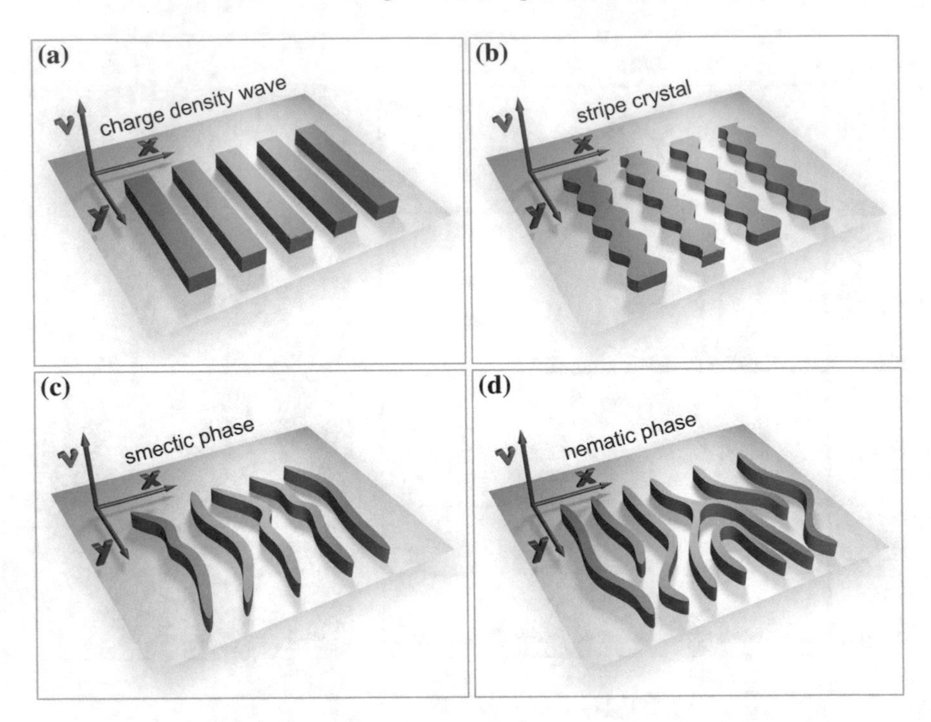


**Fig. 5.2** Illustration of the filling factor variation in the topmost Landau level for the different stripe phase models. **a** The charge density wave picture assumes alternating stripes with a strictly 1D periodicity. **b** For the stripe crystal phase, modulations on neighboring stripes are ordered anti-phase. **c** The smectic phase has random shape fluctuations with a moderate amplitude. **d** In the case of the nematic phase, the shape fluctuations are strong enough that stripes break apart and dislocations form

details remain elusive. A first step towards a microscopic understanding has been made by Kukushkin et al. [15]. They probed the collective modes and periodicity of the stripe phase at $\nu = 9/2$ by means of surface acoustic waves and photoluminescence spectroscopy. In principle, scanning probe techniques would serve as an ideal method to study the local nature of the anisotropic phase. However, their implementation is impeded by the necessity for low temperatures and the depth of the 2DES in GaAs/AlGaAs heterostructures [16]. The latter is problematic since the distance between probing tip and 2DES sets a lower bound to the spatial resolution achievable with scanning probe techniques. The density modulations in stripe phases are predicted to occur on the order of a few magnetic lengths—a length scale much smaller than the depth of a 2DES in a typical heterostructure.

In this chapter, we adapt the previously introduced NMR technique to study the microscopic nature of the anisotropic phase emerging at $\nu = 5/2$ when tilting the sample with respect to the external magnetic field. In a sense, we use nuclear spins as local detectors to probe the spatial density distribution of the anisotropic phase.

## 5.2 NMR Spectroscopy of the Stripe Phase

Following the sequence in Fig. 4.3, several nuclear resonances were measured in the filling factor range displaying large transport anisotropy at a fixed magnetic field $B_{\text{tot}} = 6.9\,\text{T}$ and a constant tilt angle of 60°. The detailed transport behavior for these conditions is shown in Fig. 5.3a. The measurement of $R_{xx}$ was performed under the influence of RF radiation in order to include off-resonant heating effects. As a consequence of the increased temperature, the $^7/_3$ and $^8/_3$ FQHS have vanished in contrast to the measurement in Fig. 5.1, which was performed at base temperature. Despite the elevated temperatures, the transport anisotropy remains on a high level. The colored bars indicate the exact filling factors at which a nuclear resonance was measured. For the detection of the NMR response, the flank of the plateau at $\nu = 3$ was chosen ($\nu_{\text{detect}} = 2.86$). The RF power was kept at a constant value of $-17\,\text{dBm}$, leading to an electron temperature of roughly 70 mK. The measured NMR spectra are presented in Fig. 5.3b. They were inverted, normalized and offset for clarity. First, we focus on the spectra recorded at $\nu = 2$ and 3. Here, two main observations leap to the eye. Firstly, the resonance at $\nu = 3$ is shifted to lower frequencies with respect to $\nu = 2$ as a result of the Knight shift. At $\nu = 3$ the electron system is maximally polarized, albeit potentially lower than $P^* = 100\,\%$ due the presence of disorder. At $\nu = 2$, in contrast, the 2DES is supposed to be unpolarized. The second observation concerns the NMR lineshape. The resonance at $\nu = 3$ is obviously broader than the

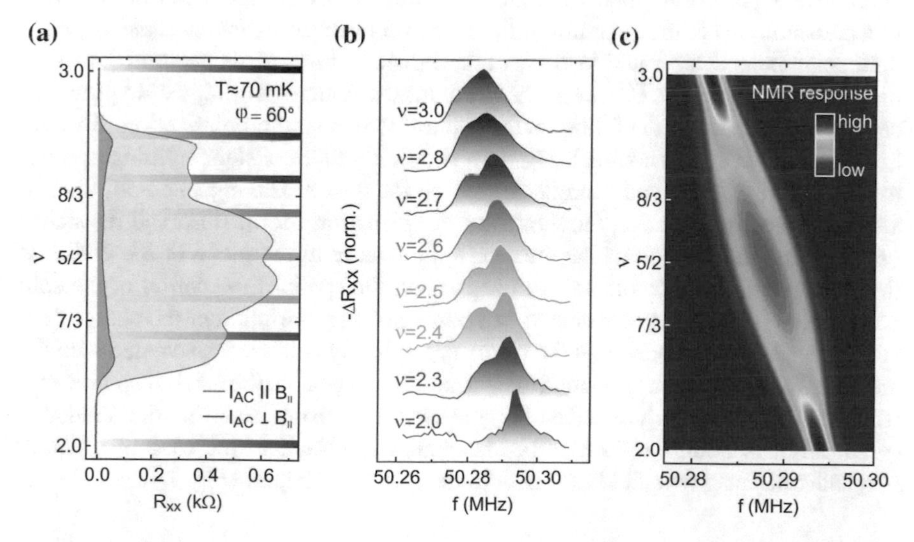


**Fig. 5.3** **a** Transport behavior along two perpendicular current directions for a constant tilt angle $\varphi = 60°$ and magnetic field $B_{\text{tot}} = 6.9\,\text{T}$. RF-induced heating caused an increase of the electron temperature to about 70 mK. **b** NMR spectra of $^{75}$As nuclei measured at different filling factors as indicated by *colored bars* in panel **a**. The resistively detected change of the nuclear spin polarization $\Delta R$ was normalized, inverted and offset for clarity. **c** Simulated NMR response based on the model in Fig. 5.4. For the calculations, a stripe period of $\lambda = 2.6 \cdot l_B$ was assumed

one at $\nu = 2$. This observation can be attributed to the finite width of the electron wavefunction as discussed in Sect. 4.2.1. The spatial integration of the varying local Knight shifts along the $z$-axis inevitably leads to a broadening of the resonance line. This effect is conspicuously absent at $\nu = 2$: In the case of an unpolarized electron system, the Knight shift is zero irrespective of the density distribution. Between these two limiting cases, at intermediate filling factors, a second resonance peak is observable in the NMR spectra. The appearance of this twofold resonance coincides with the filling factor range exhibiting a strong transport anisotropy. When increasing the filling factor, both resonances shift to higher frequencies and eventually evolve back to a single resonance at $\nu = 3$. The steady shift to lower frequencies reflects the rising degree of spin polarization when populating the spin-up branch of the second Landau level.

It is tempting to attribute the splitting of the resonance line to a modulation of the electron spin density in the plane of the 2DES. In fact, other mechanisms known to cause additional features in the NMR response can be excluded as described below. The first candidate to think of is the electric quadrupole interaction. Samples with an intrinsic electric field gradient, caused for instance by internal or external stress, exhibit multiple, evenly spaced resonances due to the electric quadrupole interaction [17, 18]. However, this effect always leads to at least three resonances in the NMR response. Moreover, any effect arising from the host crystal should be independent of the filling factor and would therefore be observable at $\nu = 2$ and 3 as well. Thus, the quadrupole interaction fails to explain the resonance splitting in Fig. 5.3b. An alternative explanation would be a modulation of the electron density in the direction perpendicular to the quantum well, caused for example by the in-plane magnetic field component. This could lead to similar NMR features if the electron density is modulated accordingly. However, the resultant resonance splitting would grow linearly when increasing the filling factor due to the rising spin polarization. It would be largest at $\nu = 3$. In contrast, Fig. 5.3b shows a rather constant splitting at intermediate filling factors and a single resonance for $\nu = 3$. Having ruled out the two alternative scenarios above, the most likely explanation for the observed resonance behavior is a modulation of the electron spin density in the plane of the 2DES. If the spin-split Landau levels are well separated, the spatial modulation of the spin polarization is inherently connected to a corresponding variation of the charge density. To further corroborate that the resonance splitting is indeed associated with the anisotropic phase, we have verified that the additional peak in the NMR response vanishes at higher temperatures when the transport anisotropy is substantially reduced (~800 mK). In addition, the twofold resonance is absent in the case of a purely perpendicular magnetic field as shown in the previous chapter (Fig. 4.4).

## 5.3 Modeling of the NMR Response

In an attempt to model the resonance behavior in Fig. 5.3b, we assume that the local filling factor is modulated in parallel stripes of $\nu_{\mathrm{loc}} = 2$ and $\nu_{\mathrm{loc}} = 3$ (Fig. 5.4a left). The relative stripe width determines the average filling factor $\nu$. The local filling factor is set by the density of the center coordinates of the electron wavefunction and might change in a step-like manner. The electron density, in contrast, is set by the probability density of the wavefunction and evolves rather smoothly due to the spatial extent of the wavefunction (Fig. 5.4a right). The exact density distribution $\rho(x, y)$ in the plane of the 2DES depends on the shape of the wavefunction and the stripe period $\lambda$. For stripes running along the $y$-axis, the density distribution is independent of the $y$-coordinate $\rho(x, y) = \rho(x)$. In a first step, the density distribution $\rho(x)$ is calculated by convolving the local filling factor $\nu_{\text{loc}}$ with the electron wavefunction $\psi$:

$$\rho_{\nu,\lambda}(x) = K \cdot \int |\psi(x - \tilde{x})|^2 \cdot (\nu_{\text{loc}}(\tilde{x}) - 2)\, d\tilde{x}, \tag{5.1}$$

using the normalization constant $K$. The calculation is done in the Landau gauge, which inherently supports the stripe symmetry (Sect. 2.3). In the second step, we determine the NMR response according to

$$\mathcal{I}(f) = \int \text{Gaussian}(f - (f_0 - \rho_{\nu,\lambda}(x) \cdot K_{s,max}))\, dx, \tag{5.2}$$

where we use again a Gaussian lineshape to describe the resonance of a single nucleus (see Eq. 4.3). Moreover, it is assumed that the spin polarization grows linearly as the spin-up branch of the second Landau level gets increasingly populated and that the detection sensitivity is uniform across the sample. The value of the maximal Knight shift $K_{s,max}$ is extracted from Fig. 5.3b. The NMR response calculated for a fixed filling factor $\nu = 5/2$ as a function of the stripe period is depicted in Fig. 5.4b. At $\nu = 5/2$, the regions with $\nu_{\text{loc}} = 2$ and $\nu_{\text{loc}} = 3$ are of equal width. The stripe period is plotted in units of the magnetic length $l_B$, which roughly sets the length scale of the wavefunction. The corresponding charge density is shown in Fig. 5.4c for prominent values of $\lambda$. We first address the two limiting cases $\lambda \gg l_B$ and $\lambda = l_B$. If the stripe period is large compared to the magnetic length, the electron density follows the step-like variation imposed by the filling factor $\nu_{\text{loc}}$. The electron system therefore alternates between unpolarized stripes and stripes having maximum polarization. Consequently, two resonances appear in the spectrum—an unshifted resonance and one with Knight shift $K_{s,max}$. In the case $\lambda = l_B$, the variation of $\nu_{\text{loc}}$ is completely smeared out by the spatial extent of the wavefunction. Hence, the spectrum exhibits only a single resonance, which is shifted by $K_{s,max}/2$ because the Landau level branch at $\nu = 5/2$ is only half occupied. At intermediate values of $\lambda$, the behavior becomes more complex. At $\lambda = 4.4 \cdot l_B$, the NMR response merges back to a single resonance. The almost uniform density distribution at this point is a result of the characteristic node appearing in the electron wavefunction of the second Landau level.

When comparing the simulated NMR response with the experimental findings in Fig. 5.3b, best agreement is found for $\lambda = (2.6 \pm 0.6) \cdot l_B$, which corresponds to $\lambda = (1.5 \pm 0.4) \cdot R_c$. For this value, both theory and experiment yield a twofold resonance which is shifted with respect to an unpolarized 2DES. For stripe phases in perpendicular magnetic fields, theory predicts a slightly higher value $\lambda_{\text{theory}} = 2.8 \cdot R_c$

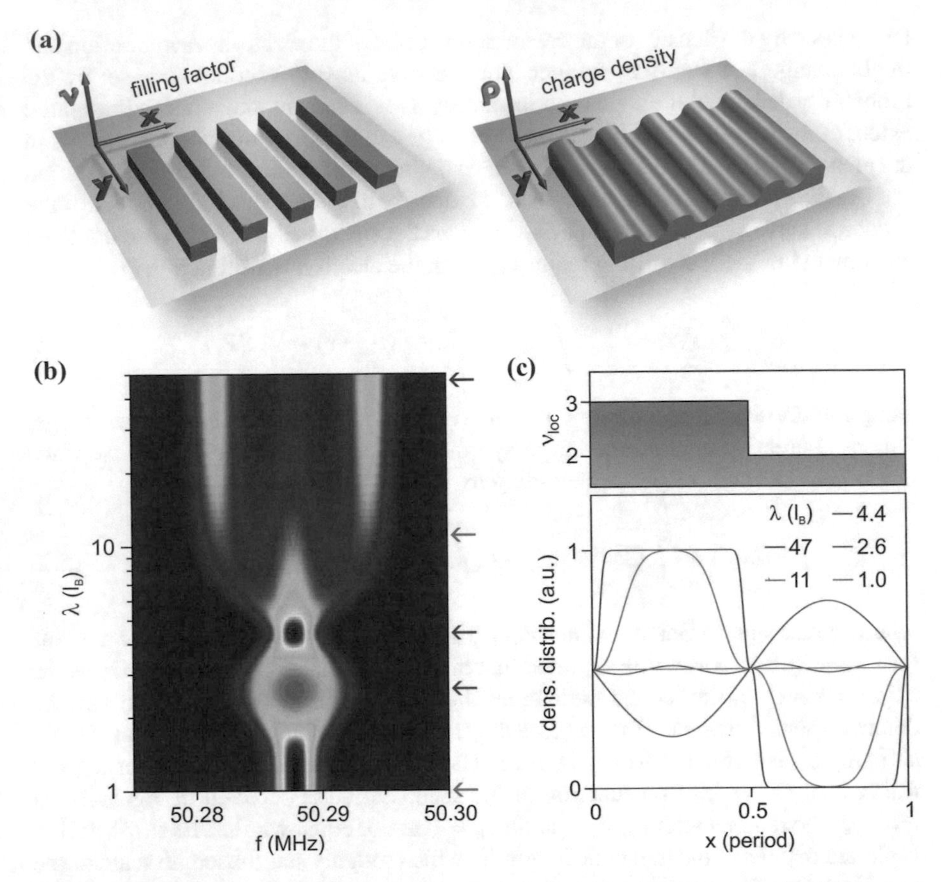


**Fig. 5.4** Modeling of the NMR response. **a** The filling factor in the topmost Landau level is assumed to alternate between $\tilde{\nu} = 1$ and 0 in parallel stripes (*left*). The spatial extent of the wavefunction leads to a smooth variation of the electron density (*right*). **b** NMR response at $\nu = 5/2$ for different stripe periods in units of the magnetic length $l_B$. **c** Spatial density distribution in the second Landau level within a single stripe period for selected points in panel **b**

[19]. No other experimentally determined values for the present conditions are known to exist. For the stripe phase at $\nu = 9/2$ in perpendicular magnetic fields, a stripe period $3.6 \cdot R_c$ was found [15].

Remarkable is also the strong charge density modulation of about 60 % (20 % of total density $n_e$) which is necessary to reproduce the resonance splitting found in experiment. Interestingly, the density modulation is ordered anti-phase with respect to the variation of the filling factor. The strong density modulation is particularly surprising considering that theoretical studies estimate a variation of only 20 % within a spin-split Landau level at zero tilt [5, 6]. To the best of our knowledge, none of the theoretical publications have addressed the effect of an in-plane magnetic field on the modulation strength. Intuitively, it comes as no surprise that the in-plane field component imparts a stronger one-dimensional character to the stripes.

Based on the extracted value $\lambda = 2.6 \cdot l_B$, we have modeled the filling factor dependence of the nuclear resonance behavior. The result is shown in Fig. 5.3c. Keeping in mind that our model is relatively straightforward and does not consider fluctuations of the stripe pattern, the simulated NMR response compares remarkably well with the experimental data in Fig. 5.3b. In both experiment and theory, the NMR spectrum starts off with a single resonance which undergoes a splitting into two distinct features. Both resonances shift simultaneously to lower frequencies when increasing $\nu$ and eventually merge back to a symmetric resonance line at $\nu = 3$. Apparent discrepancies mainly arise from the fact that our model is purely two-dimensional. Calculating the density distribution in the $z$-direction as done in the previous chapter is in the present case rather cumbersome due to the influence of the in-plane magnetic field component on the shape of the electron wavefunction. This task needs to be addressed in future work. A three-dimensional model would capture the broadening of the resonances occurring at higher filling factors. Presumably also the asymmetric height of the resonances in Fig. 5.3b, with the resonance at lower frequencies being constantly smaller, is a consequence of the electron distribution in the $z$-direction. As pointed out in Sect. 4.2.1, the spatial spread of the wavefunction in the direction perpendicular to the quantum well often causes an asymmetric lineshape slanted towards higher frequencies.

From the clear splitting of the resonance line in Fig. 5.3b, one may further conjecture that the stripe pattern is rather well ordered. Strong fluctuations along the stripes would change the stripe period locally, resulting in a different NMR response according to Fig. 5.4b. This would smooth out and obscure the twofold resonance when averaging across the whole sample. Hence, the appearance of two distinct resonances points towards the formation of a strongly ordered phase, such as a unidirectional CDW or a stripe crystal. This ordered phase is presumably stabilized by the in-plane field component. It has been shown theoretically that an in-plane magnetic field enhances the anisotropy energy of the stripe phase [20]. The increased stability of the stripe phase under tilt also manifests itself in the need for higher temperatures to destroy the transport anisotropy [12].

It is noteworthy that a twofold nuclear resonance similar to the one discussed here has recently been discovered in the high-$T_c$ superconductor $YBa_2Cu_3O_y$. There as well, the additional resonance peak was interpreted as evidence for a magnetic-field-induced charge-stripe order [21].

### Outlook

Future work will aim at repeating the above experiments for the stripe phases in higher Landau levels. This would allow comparing the different stripe phases among each other. Experimentally it is challenging to acquire resonance spectra at higher Landau levels due to the lower magnetic fields. The reduced field has two negative implications for performing NMR experiments. Firstly, it decreases the nuclear spin polarization and therefore the signal strength. Secondly, the Knight shift is reduced as well owing to the lower Landau level degeneracy. This impairs the chance to detect single features in the NMR response. In these regards, the stripe phase at $\nu = 5/2$ provides the best conditions for NMR spectroscopy. Here, the Landau level degeneracy set be the perpendicular magnetic field component is comparably large, and the total magnetic field, relevant for the degree of nuclear spin polarization, is much higher due to the tilted field configuration. Apart from that, the electron wavefunction becomes increasingly complicated in higher Landau levels (see Fig. 2.7). Consequently, the spatial density distribution of the stripe phase and the NMR response would presumably be more complex as well.

Another project currently under investigation employs NMR spectroscopy to probe the density distribution of the different bubble phases in the quantum Hall regime. This is particularly challenging since the reentrant states in the second Landau level are extremely fragile. Tilting the sample to increase the NMR signal is here not an option as the bubble phases immediately disappear under the influence of an in-plane magnetic field component. Yet, a close relative of the bubble phase, the Wigner crystal, has recently been successfully analyzed using the NMR technique employed above [22].

## References

1. J.P. Eisenstein, R. Willett, H.L. Stormer, D.C. Tsui, A.C. Gossard, J.H. English, Collapse of the even-denominator fractional quantum Hall effect in tilted fields. Phys. Rev. Lett. **61**, 997 (1988)
2. M.P. Lilly, K.B. Cooper, J.P. Eisenstein, L.N. Pfeiffer, K.W. West, Anisotropic states of two-dimensional electron systems in high Landau levels: Effect of an in-plane magnetic field. Phys. Rev. Lett. **83**, 824 (1999)
3. W. Pan, R.R. Du, H.L. Stormer, D.C. Tsui, L.N. Pfeiffer, K.W. Baldwin, K.W. West, Strongly anisotropic electronic transport at Landau level filling factor $\nu = 9/2$ and $\nu = 5/2$ under a tilted magnetic field. Phys. Rev. Lett. **83**, 820 (1999)
4. J. Xia, V. Cvicek, J.P. Eisenstein, L.N. Pfeiffer, K.W. West, Tilt-induced anisotropic to isotropic phase transition at $\nu = 5/2$. Phys. Rev. Lett. **105**, 176807 (2010)
5. A.A. Koulakov, M.M. Fogler, B.I. Shklovskii, Charge density wave in two-dimensional electron liquid in weak magnetic field. Phys. Rev. Lett. **76**, 499 (1996)
6. M.M. Fogler, A.A. Koulakov, B.I. Shklovskii, Ground state of a two-dimensional electron liquid in a weak magnetic field. Phys. Rev. B **54**, 1853 (1996)
7. R. Moessner, J.T. Chalker, Exact results for interacting electrons in high Landau levels. Phys. Rev. B **54**, 5006 (1996)
8. E. Fradkin, S.A. Kivelson, Liquid-crystal phases of quantum Hall systems. Phys. Rev. B **59**, 8065 (1999)
9. M.M. Fogler, Stripe and bubble phases in quantum Hall systems, in *High Magnetic Fields: Applications in Condensed Matter Physics and Spectroscopy* (Springer, Berlin, 2002), pp. 98–138
10. H.A. Fertig, Unlocking transition for modulated surfaces and quantum Hall stripes. Phys. Rev. Lett. **82**, 3693 (1999)
11. E. Fradkin, S.A. Kivelson, E. Manousakis, K. Nho, Nematic phase of the two-dimensional electron gas in a magnetic field. Phys. Rev. Lett. **84**, 1982 (2000)
12. K.B. Cooper, M.P. Lilly, J.P. Eisenstein, L.N. Pfeiffer, K.W. West, Onset of anisotropic transport of two-dimensional electrons in high Landau levels: Possible isotropic-to-nematic liquid-crystal phase transition. Phys. Rev. B **65**, 241313 (2002)
13. G. Sambandamurthy, R.M. Lewis, H. Zhu, Y.P. Chen, L.W. Engel, D.C. Tsui, L.N. Pfeiffer, K.W. West, Observation of pinning mode of stripe phases of 2D systems in high Landau levels. Phys. Rev. Lett. **100**, 256801 (2008)

14. H. Zhu, G. Sambandamurthy, L.W. Engel, D.C. Tsui, L.N. Pfeiffer, K.W. West, Pinning mode resonances of 2D electron stripe phases: Effect of an in-plane magnetic field. Phys. Rev. Lett. **102**, 136804 (2009)
15. I.V. Kukushkin, V. Umansky, K. von Klitzing, J.H. Smet, Collective modes and the periodicity of quantum Hall stripes. Phys. Rev. Lett. **106**, 206804 (2011)
16. A. Yacoby, H.F. Hess, T.A. Fulton, L.N. Pfeiffer, K.W. West, Electrical imaging of the quantum Hall state. Solid State Commun. **111**, 1 (1999)
17. J.H. Smet, R.A. Deutschmann, W. Wegscheider, G. Abstreiter, K. von Klitzing, Ising ferromagnetism and domain morphology in the fractional quantum Hall regime. Phys. Rev. Lett. **86**, 2412 (2001)
18. W. Desrat, D.K. Maude, M. Potemski, J.C. Portal, Z.R. Wasilewski, G. Hill, Resistively detected nuclear magnetic resonance in the quantum Hall regime: Possible evidence for a Skyrme crystal. Phys. Rev. Lett. **88**, 256807 (2002)
19. M.O. Goerbig, P. Lederer, C. Morais Smith, Competition between quantum-liquid and electron-solid phases in intermediate Landau levels. Phys. Rev. B **69**, 115327 (2004)
20. T. Jungwirth, A.H. MacDonald, L. Smrčka, S.M. Girvin, Field-tilt anisotropy energy in quantum Hall stripe states. Phys. Rev. B **60**, 15574 (1999)
21. T. Wu, H. Mayaffre, S. Krämer, M. Horvatić, C. Berthier, W.N. Hardy, R. Liang, D.A. Bonn, M.-H. Julien, Magnetic-field-induced charge-stripe order in the high-temperature superconductor $YBa_2Cu_3O_y$. Nature **477**, 191 (2011)
22. L. Tiemann, T.D. Rhone, N. Shibata, K. Muraki, NMR profiling of quantum electron solids in high magnetic fields. Nat. Phys. **10**, 648 (2014)

# Chapter 6
# Surface Acoustic Wave Study of Density-Modulated Phases

The existence of density-modulated phases was first inferred from signatures in charge transport experiments [1, 2]. The bubble phase was identified by a reappearance of the integer quantum Hall effect, whereas the stripe phase manifested itself by the anisotropic transport behavior (Sect. 2.5). Since then, research on density-modulated phases has focused primarily on their response to an external current flow. Here, we pursue a different approach and investigate these phases by means of surface acoustic waves. This technique provides an intriguing alternative access to the electron dynamics and conductivity in a two-dimensional electron system.

## 6.1 Surface Acoustic Waves on a GaAs/AlGaAs Heterostructure

This section summarizes the basic properties of surface acoustic waves and discusses their interaction with a two-dimensional electron system. The information provided here is based on references [3, 4].

Surface acoustic waves (SAWs) are modes of elastic energy bound to propagate along the surface of an elastic medium [3]. They combine elements of longitudinal and transversal waves. Hence, the displacement of a single particle is elliptic. Their amplitude decays exponentially towards the interior of the sample at a length scale set by the SAW wavelength. In piezoelectric materials, such as GaAs/AlGaAs, the mechanical strain associated with the particle displacement creates an electric field, which propagates alongside the sound wave. The presence of this field significantly alters the SAW propagation in the material. The piezoelectric tensor $\mathfrak{p}$ couples the mechanical and electrical quantities according to

$$\mathcal{T}_{i,j} = c_{ijkl}\frac{\partial u_k}{\partial \mathrm{x}_l} + \mathfrak{p}_{kij}\frac{\partial \phi}{\partial \mathrm{x}_k} \tag{6.1}$$



B. Frieß, *Spin and Charge Ordering in the Quantum Hall Regime*,
Springer Theses, DOI 10.1007/978-3-319-33536-0_6

$$D_i = -\varepsilon_{ij}\frac{\partial \phi}{\partial \mathrm{x}_j} + \mathfrak{p}_{ijk}\frac{\partial u_j}{\partial \mathrm{x}_k}, \tag{6.2}$$

with indices $i, j, k, l = 1, 2, 3$ [4]. The first term in the stress equation (Eq. 6.1) represents Hooke's law with the particle displacement $u$, stiffness tensor $c$ and mechanical stress $\mathcal{T}$. The spatial coordinates are denoted as $x_i$. The first term of the electric equation (Eq. 6.2) is the well-known relation between electric field $\boldsymbol{E}$ and displacement field $\boldsymbol{D} = \epsilon \boldsymbol{E}$, where $\epsilon$ is the electric permittivity tensor and $\phi$ the scalar electric potential.

To simplify the equations above, we first consider a one-dimensional longitudinal bulk wave running along the $x$-direction. Combining

$$\frac{\partial \mathcal{T}}{\partial x} = \rho\frac{\partial^2 u}{\partial t^2} \tag{6.3}$$

with Eqs. 6.1 and 6.2 leads to the basic equation of motion

$$\rho\frac{\partial^2 u}{\partial t^2} = c\left(1 + \frac{\mathfrak{p}^2}{c\,\epsilon}\right)\frac{\partial^2 u}{\partial x^2} - \frac{\mathfrak{p}}{\epsilon}\frac{\partial D}{\partial x}, \tag{6.4}$$

where $\rho$ is the density of the material [3]. Based on this equation, we discuss the two opposing scenarios of an insulating and a perfectly conducting piezoelectric material.

- In the case of an ideal conductor ($\sigma = \infty$), any electric field generated by the SAW is immediately screened by the electrons provided that the electronic relaxation time, i.e. the time required by the electrons to react to an external electric field, is shorter than $1/f$ (SAW frequency $f$). As a consequence, the last terms in Eq. 6.4 cancel each other, and it remains

$$\rho\frac{\partial^2 u}{\partial t^2} = c\frac{\partial^2 u}{\partial x^2}. \tag{6.5}$$

  This is the equation of a wave propagating with velocity $v_0 = \sqrt{c/\rho}$. Hence, the material behaves as if it would be non-piezoelectric.
- In the case of a piezoelectric insulator ($\sigma = 0$), no charges can build up to screen the electric field, and according to Poisson's equation, $\partial D/\partial x = 0$ follows. Consequently, Eq. 6.4 describes waves propagating with velocity $v = \sqrt{\tilde{c}/\rho}$ in a material with an effective stiffness $\tilde{c} = c(1 + \mathfrak{p}^2/c\,\epsilon)$. In this case, the piezoelectricity leads to a stiffening of the crystal. The term $\mathfrak{p}^2/c\,\epsilon$ is also known as the electromechanical coupling coefficient $\mathrm{K}^2$.

In summary, the extent to which the SAW electric fields are screened depends decisively on the conductivity $\sigma$ as well as the SAW frequency $\omega$. In this context, it is instructive to introduce the conductivity relaxation frequency $\omega_r = \sigma/(\epsilon_1 + \epsilon_2)$ [3]. This frequency is derived from the time $t = 2\pi/\omega_r$ which an electron system needs to react to an external electric field and to restore equilibrium. Here, $\epsilon_1$ and $\epsilon_2$ denote the dielectric constants of the sample and the space above, respectively. If $\omega < \omega_r$, the electrons are able to follow the electric field created by the SAW and build up a space charge which screens the electric field. If in contrast $\omega > \omega_r$, the electrons cannot redistribute fast enough, and the SAW propagates like in a piezoelectric insulator. In this case, a piezoelectric stiffening of the crystal occurs. For the general case of SAW propagation on top of a piezoelectric material with homogeneous bulk conductivity $\sigma$, the velocity change $\Delta v/v_0$ and the attenuation coefficient $\Gamma$ can be expressed as

$$\frac{\Delta v}{v_0} = \frac{\mathrm{K}_{\mathrm{eff}}^2}{2} \frac{1}{1+(\frac{\omega_r}{\omega})^2} \tag{6.6}$$

$$\Gamma = \frac{\omega}{v_0} \frac{\mathrm{K}_{\mathrm{eff}}^2}{2} \frac{\omega_r/\omega}{1+(\frac{\omega_r}{\omega})^2}. \tag{6.7}$$

Here, $\mathrm{K}_{\mathrm{eff}}$ denotes the effective coupling coefficient which takes into account the boundary conditions on the surface of the material [3, 5].

Yet, the equations above do not fully reflect the physical situation in GaAs/AlGaAs heterostructures since here the bulk is insulating (at low temperatures) and the electrons are confined to a thin conductive layer. This case has been approached theoretically by considering a metallic layer with sheet conductivity $\sigma_\square$ on the surface of a piezoelectric insulator. It was found that under these circumstances the relaxation frequency $\omega_r$ depends on the SAW wavevector $k$ as [3, 6]

$$\omega_r = \frac{\sigma_\square k}{\epsilon_1 + \epsilon_2}. \tag{6.8}$$

Consequently, Eqs. 6.6 and 6.7 can be expressed in terms of conductivities

$$\frac{\Delta v}{v_0} = \frac{\mathrm{K}_{\mathrm{eff}}^2}{2} \frac{1}{1+(\frac{\sigma_\square}{\sigma_\mathrm{m}})^2} \tag{6.9}$$

$$\Gamma = k\frac{\mathrm{K}_{\mathrm{eff}}^2}{2} \frac{\sigma_\square/\sigma_\mathrm{m}}{1+(\frac{\sigma_\square}{\sigma_\mathrm{m}})^2}, \tag{6.10}$$

where $\sigma_\mathrm{m} = v_0(\epsilon_1 + \epsilon_2)$ was used. The velocity change $\Delta v/v_0$ and SAW attenuation $\Gamma/k$ for a GaAs (100) surface are plotted in Fig. 6.1a, b, respectively. It was assumed that $\sigma_\mathrm{m} = 3.3 \times 10^{-7} (\Omega/\square)^{-1}$ and $\mathrm{K}_{\mathrm{eff}}^2 = 6.4 \times 10^{-4}$ [3]. For conductivities above $\sigma_\mathrm{m}$, the SAW velocity is identical to $v_0$. In the opposite case, the relative velocity change equals $\mathrm{K}_{\mathrm{eff}}^2/2$. The SAW attenuation results from ohmic losses in the material and becomes strongest for $\sigma_\square = \sigma_\mathrm{m}$. It drops again for lower conductivities because here current and electric field are out of phase [5].

If we now consider looseness-1the propagation of surface acoustic waves in an external magnetic field, the sheet conductivity $\sigma_\square$ in the equations above has to be replaced by the longitudinal conductivity $\sigma_{xx}$, assuming that the electric field generated by the SAW is oriented along the $x$-direction [3]. Surface acoustic waves

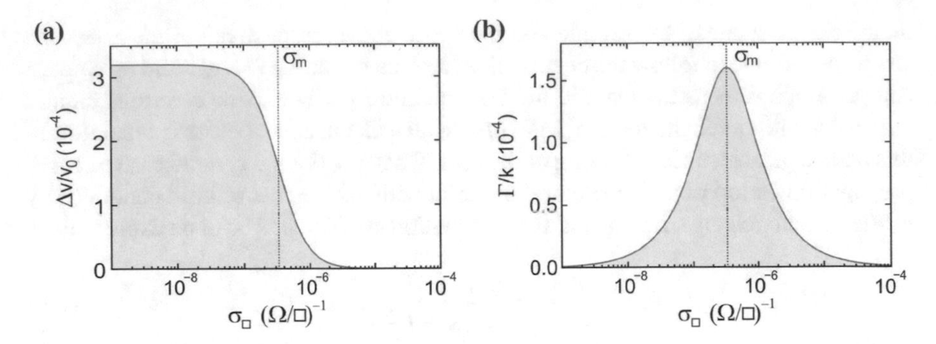


**Fig. 6.1** Velocity change (**a**) and damping factor (**b**) calculated for a SAW propagation along the GaAs (100) surface. For the calculations, the parameters $\sigma_m = 3.3 \times 10^{-7}(\Omega/\square)^{-1}$ and $K^2_{eff} = 6.4 \times 10^{-4}$ were used

are well suited to study the conductivity oscillations in the quantum Hall regime if the depth of the 2DES is considerably smaller than the SAW wavelength. Figure 6.1 tells us that the SAW propagation is primarily sensitive to very low conductivities. Hence, the velocity change is strongest in the region of incompressible quantum Hall states. In fact, the model described above was found to successfully reproduce the SAW propagation in the quantum Hall regime [7].

The study of quantum Hall physics using surface acoustic waves offers intriguing insights also beyond the basic transport behavior. Presumably the most prominent example is the evidence for composite fermions (Sect. 2.4.2) found by Willett et al. when detecting a Fermi surface at half-filled Landau levels [8–10]. In detail, an enhanced conductivity was observed at $\nu = 1/2$ if the SAW wavelength is tuned below the mean free path of composite fermions. When setting the magnetic field slightly away from $\nu = 1/2$, geometrical resonances appeared in the conductivity, corresponding to the cyclotron motion of the composite fermions in their reduced effective magnetic field.

SAW studies are advantageous in many regards. Not only that surface acoustic waves probe the conductivity of a 2DES at dimensions set by the SAW wavelength, they conveniently do so without the need for electrical contacts to the sample. In addition, the probing direction of the acoustic waves is well defined in contrast to macroscopic transport measurements. Especially in the case of the density-modulated phases, the direction of current flow is strongly influenced by the spatially varying density distribution. In the stripe phase, for instance, the current spreads towards the edges of the sample when being sent perpendicular to the stripe orientation [11, 12]. Surface acoustic waves allow probing the inner parts of the sample even in this case. We will benefit from this property of SAWs in the following sections.

## 6.2 Experimental Setup and Measurement Technique

To study the propagation of surface acoustic waves in the quantum Hall regime, we fabricated the sample structure depicted in Fig. 6.2. As in the previous chapter, the 2DES is confined to a square-shaped mesa structure in order to obtain an equal probing length along the [110] and $[1\bar{1}0]$ crystal directions. On either side of the sample, a so-called interdigital transducer is placed. It is made of aluminum with a thickness of 100 nm. Each transducer consists of two electrodes with an interlaced comb-like structure. These interdigital transducers are used to excite surface acoustic waves in the GaAs/AlGaAs heterostructure. For this purpose, a high-frequency voltage is applied between both electrodes of the transducer. As a result, an electric field spans the gap between the finger structure and penetrates the substrate below. Here, the piezoelectricity causes an alternating density variation in response to the applied microwave field. These density oscillations excite surface acoustic waves, which propagate along the surface of the substrate. The wavelength of the SAW is set by the period of the comb-shaped electrodes. After traversing the mesa structure, the SAW arrives at the opposite transducer and gets partially converted back into an electrical signal. While the surface wave is moving across the mesa, the 2DES underneath is exposed to the alternating electric field and therefore tries to screen it by redistributing the electrons. Depending on the conductivity of the 2DES, the screening will be more or less effective. If the electrons successfully weaken the electric field, the

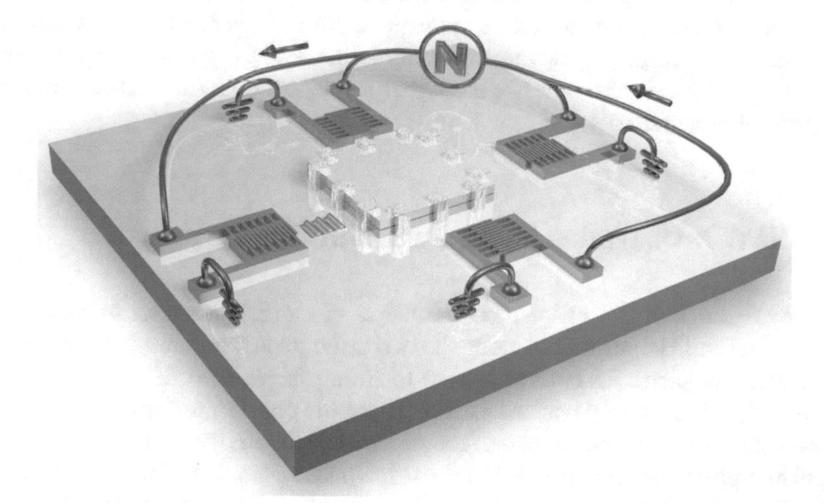


**Fig. 6.2** Schematic of the sample used for the SAW experiments (not drawn to scale). The 2DES is confined to a *square-shaped* mesa (width 1.1 mm). On either side of the mesa, an interdigital transducer was fabricated out of aluminum. In the $[1\bar{1}0]$ direction, the pair of transducers was placed further apart to distinguish the SAW propagation along both crystal directions by their travel time. The SAW phase shift and power transmission was measured using a standard network analyzer (represented by *blue* "*N*")

SAW is slowed down as discussed in the previous section (Fig. 6.1a). This velocity change is detectable as a phase shift of the SAW when reaching the transducer at the opposite side of the structure. Given the phase difference $\Delta\alpha$, the velocity change $\Delta v/v_0$ can be calculated according to

$$\frac{\Delta v}{v_0} = \frac{\lambda}{L}\frac{\Delta\alpha}{360^\circ}, \tag{6.11}$$

where $\lambda$ is the SAW wavelength and $L$ the length of the mesa. The reference point $v_0$ for the velocity change $\Delta v$ is standardly chosen such that it resembles the case of a 2DES with perfect screening properties, i.e. when no piezoelectric stiffening of the crystal occurs.

A technical subtlety which had to be dealt with is the constraint that only two coaxial cables are fitted to the microwave sample holder, owing to the endeavor to minimize the heat load. Yet, to be able to probe the 2DES independently along the two major axes with only one pair of transmission lines, the signals emanating from both directions need to be distinguished. This was accomplished by choosing different travel times of the surface acoustic waves along each axis. For this purpose, one set of transducers was placed further apart from the mesa. The two axes can then be addressed individually by gating the transmission time. For the present design, the traveling times are between 600 ns and 1.4 µs. A network analyzer was utilized to apply the microwave signal as well as to detect the response of the second, opposite transducer. Both the phase and transmitted power were measured simultaneously. The signal of the incoming surface acoustic wave is in general rather small, and often multiple averaging cycles are necessary to improve the signal quality. It turned out to be crucial to further enhance the signal-to-noise ratio by an external amplifier (amplification factor $10^4$).

## 6.3 SAW Propagation in the Quantum Hall Regime

The key findings of this chapter are presented in Fig. 6.3. It shows in panel a the longitudinal and Hall resistance along the two orthogonal directions [110] and [$1\bar{1}0$] at different magnetic fields from $B = 0$ to filling factor $\nu = 2$. Between $\nu = 4$ and $\nu = 8$, different stripe phases as well as bubble phases are discernible by their characteristic transport behavior—a reentrance of the IQHE for the bubble phase and an anisotropic transport in the case of the stripe phase (Sect. 2.5). Panel b depicts the phase shift and velocity change of the surface acoustic waves propagating along both crystal directions with a frequency of 340 MHz. The power transmitted by the SAW is shown in panel c. The data below 150 mT is omitted because the aluminum of the transducers turns superconducting at low magnetic fields. This transition also affects the SAW propagation. The main observations of Fig. 6.3 are summarized below.

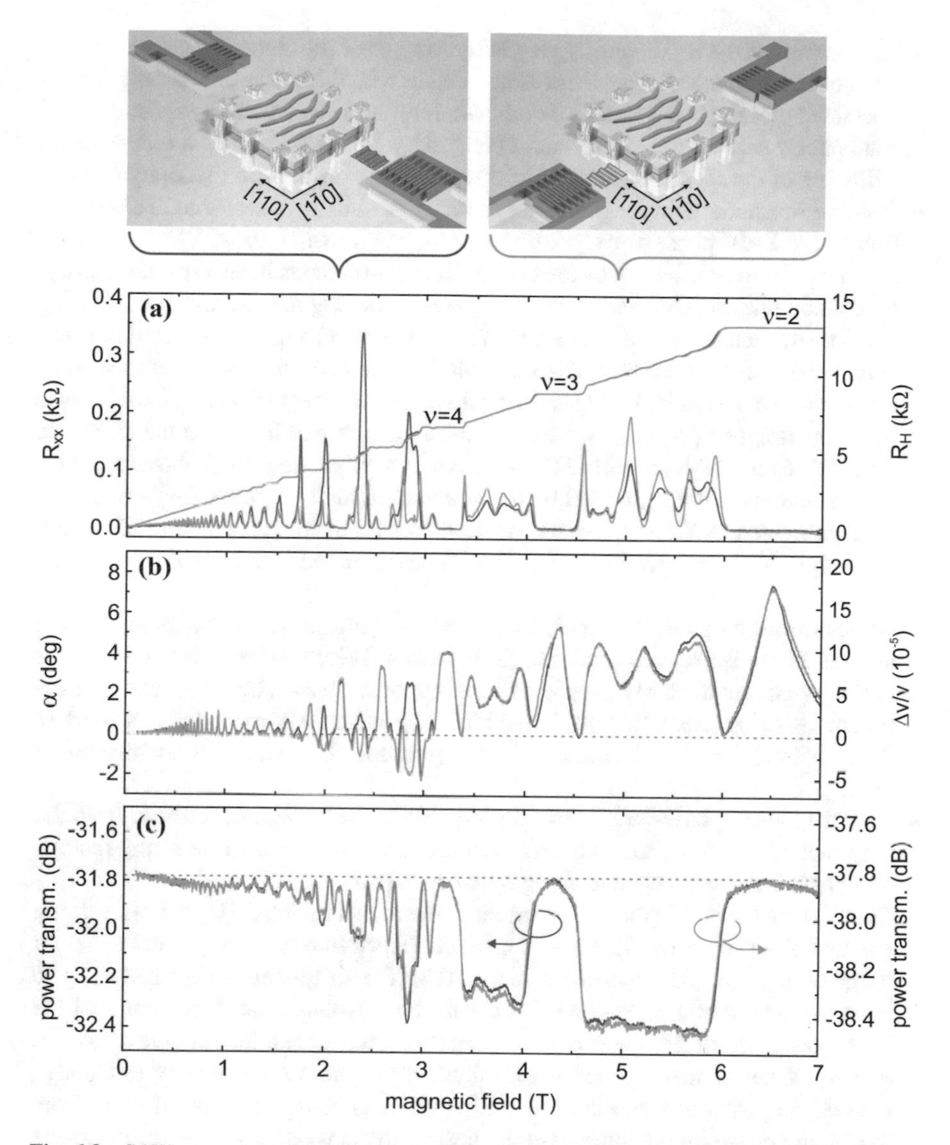


**Fig. 6.3** SAW propagation in the quantum Hall regime. **a** Longitudinal and Hall resistance as a function of magnetic field. The current was sent along the perpendicular directions [110] (*black*) and [1$\bar{1}$0] (*blue*). *Green stripes* on top of the depicted mesa indicate the stripe orientation as deduced from the transport behavior. **b** Phase difference and velocity change for SAW propagation along both crystal directions. $\alpha(B = 0)$ was chosen as zero point. **c** SAW transmission in the same magnetic field range. In the [1$\bar{1}$0] direction the SAW transmission is lower due to the higher spatial separation of the transducers. The transmission is increased by an external amplifier (factor $10^4$)

- An increase of the SAW velocity is visible whenever the electron system enters a quantum Hall state. This behavior is consistent with the theoretical predictions of the model presented above: The incompressible electron system in a quantum Hall state poorly screens the electric field created by the SAW. Hence, a piezoelectric stiffening of the crystal occurs, which increases the propagation velocity.
- In some magnetic field ranges, a negative phase shift is observable. This reduction of the SAW phase corresponds to a slowing down of the SAW velocity or, alternatively, to a softening of the crystal. Its occurrence is limited to the density-modulated phases. The discovery of a crystal softening is surprising considering that for any conductivity a positive (or zero) velocity change is predicted to occur (Fig. 6.1a). In fact, to the best of our knowledge, this is the first observation of such a behavior. However, it all hinges on the question whether the zero point reference has been assigned properly. We have chosen the phase at $B = 0$ as the reference point since for a high mobility 2DES the conductivity here is well above $\sigma_m$. This assignment is further supported by the behavior at high magnetic fields, where $\alpha$ approaches zero multiple times whenever the 2DES becomes compressible again. The origin of the negative phase shift will be discussed in more detail at a later point.
- Another intriguing observation is made for the stripe phases in the quantum Hall regime. The anisotropy observed so far in standard electron transport experiments is also evident in the SAW propagation. For acoustic waves traveling in the direction of the stripe orientation (as derived from $R_{xx}$), the SAW velocity is reduced. In the perpendicular case, the acoustic waves gain speed, leading to a positive phase shift.
- Also the bubble phases are clearly discernable in the SAW propagation. Both the longitudinal resistance and the SAW velocity are isotropic at these filling factors. The phase change is negative along both crystal directions.
- Considering the SAW power transmission, the model in Fig. 6.1b predicts a strong attenuation of the acoustic waves for low conductivities $\sigma \approx \sigma_m$. This behavior is reproduced by the experiment in Fig. 6.3c for all quantum Hall states up to a magnetic field of 1 T. Beyond that point, the magnetic field dependence of the SAW power inverts. It stands to reason that this behavior results from the crossover $\sigma < \sigma_m$. When comparing the transmission intensity among the density-modulated phases, the power absorption of the stripe phase is conspicuously different from the bubble phase. In the latter case the absorption is weak, whereas for the stripe phase a strong damping is observed. Interestingly, the attenuation in the stripe phase is equal along both crystal directions. The overall SAW transmission is reduced along the $[1\bar{1}0]$ direction owing to the increased spatial separation between the transducers on this axis.

Before embarking on the discussion of the observed SAW transmission, we briefly look at the temperature dependence in Fig. 6.4. Here, the measurements of Fig. 6.3 were repeated at higher temperatures (67, 78 and 89 mK). With increasing temperature, corresponding features in $R_{xx}$ and $\alpha$ disappear concomitantly starting with the less robust phases at low magnetic fields. Figure 6.4 emphasizes the inherent

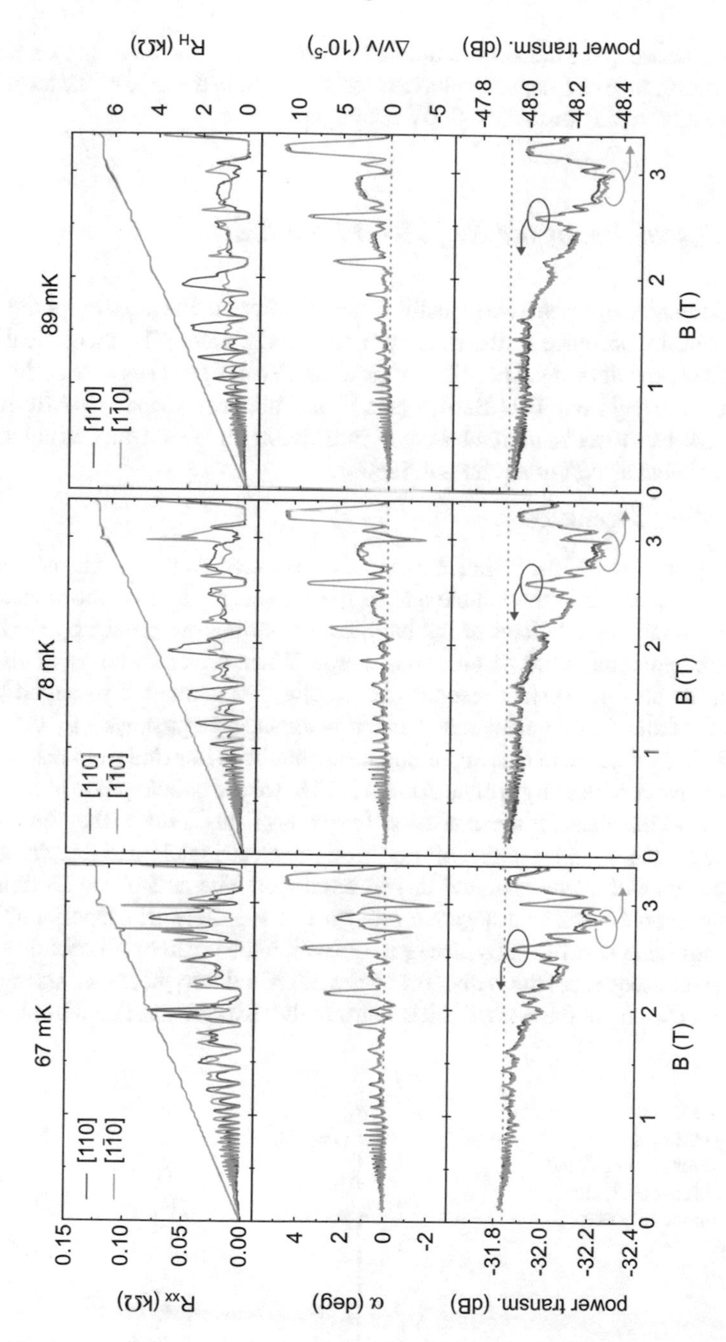


**Fig. 6.4** Temperature dependence of the transport measurements in Fig. 6.3. The studied temperatures are 67 mK (*left column*), 78 mK (*central column*) and 89 mK (*right column*). The *black* (*blue*) line indicates the current flow and SAW propagation in the [110] ([1$\bar{1}$0]) direction. All features in $R_{xx}$ and $\alpha$ which are associated with the appearance of density-modulated phases exhibit equal temperature dependencies

connection between the density-modulated phases on the one hand and on the other hand the occurrence of a negative phase shift as well as in the case of the stripe phase the observation of an anisotropic SAW propagation.

### *6.3.1 Discussion of the Negative Phase Shift*

In the following, we present two plausible interpretations of the negative phase change occurring in the presence of the density-modulated phases. The theoretical understanding has been developed in collaboration with Prof. Bernd Rosenow (Universität Leipzig), Dr. Yang Peng (FU Berlin) and Prof. Felix von Oppen (FU Berlin). We restrict ourselves to a discussion based on intuitive arguments. Detailed calculations will be available as part of a joint publication.

**Pinning Mode Resonance**

A first interpretation of the negative phase shift is based on the occurrence of a pinning mode resonance in the bubble and stripe phases. In the realistic scenario of a disordered sample, the lattices of the bubbles and stripes are pinned by the random electrostatic potential of the disorder landscape. When driven by an oscillating electric field, the electron system responds in a collective manner, a so-called pinning mode, if the external frequency matches the resonance frequency set by the pinning potential. The existence of such a pinning mode has been inferred from resonances in the microwave conductivity around 200 MHz [13] for the bubble phase and 100 MHz [14] for the stripe phase. If the excitation frequency is higher than the pinning mode resonance frequency, a $\pi$ phase shift between the electric field and the charge modulation occurs as for any classical driven oscillator. The anti-phase motion of the oscillating charge leads to a negative real part of the dielectric function (Fig. 6.5) whose magnitude is sufficiently strong to outweigh the positive contributions of the bulk. As a consequence, the reduction of the SAW velocity becomes stronger than for a sole screening of the SAW electric field. In the stripe phase, the pinning strength

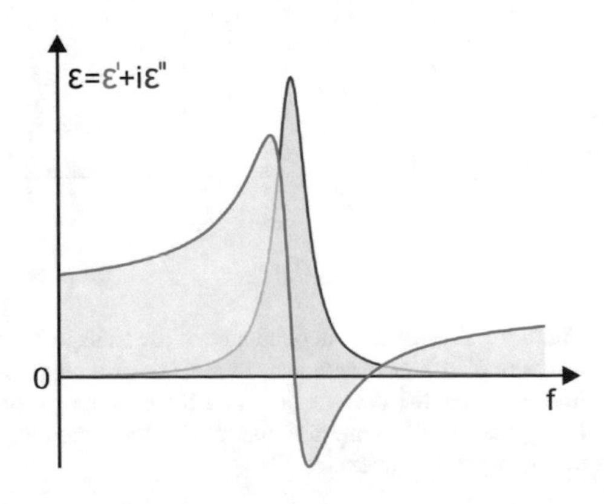


**Fig. 6.5** Real and imaginary part of the dielectric function calculated from the Lorentz oscillator model in the vicinity of a resonance

depends on the crystal direction. Along the stripe orientation, the pinning is small or completely absent in strong contrast to the perpendicular direction. Hence, the anisotropic SAW propagation in the stripe phase arises partly from the anisotropic nature of the pinning mode resonance. However, from the above sequence of arguments, one would infer that a negative velocity change only occurs if the acoustic waves travel along the easy direction, contrary to the observations in Fig. 6.3b. The missing piece of information is the contribution of the magnetic field. It couples the equations of motion for the two orthogonal directions and ultimately causes an inversion of the anisotropy.

The observation of an identical SAW phase shift in both crystal directions of the bubble phases suggests that the pinning mode is isotropic here. In an advanced interpretation, this points to the symmetric Wigner crystal as the basis for the bubble phase. Anisotropic models, such as a pinned stripe crystal (Fig. 5.2), seem to be in conflict with the isotropic sound propagation. So far, it has not been possible to address the degree of anisotropy in the bubble phase by macroscopic transport experiments due to its insulating nature.

Concerning the SAW power transmission, the energy dissipation is expected to be highest at the pinning mode resonance. At larger frequencies, the dissipation is supposed to be small since here the current and electric field are out of phase as for any driven harmonic oscillator. This prediction fits the experimental findings in the case of the bubble phase. For the stripe phase, however, a low power transmission (high absorption) is found in experiment. This contrast is better explained by the different compressibilities of the respective phases. In the bubble phase, the electron system is incompressible similar to the nearby IQHS. Hence, both states exhibit minimal power absorption due to their low bulk conductivities. In case of the stripe phase, the situation is reversed: The electron system is compressible and has a moderate bulk conductivity. Consequently, the power absorption is enhanced, similar to other close-by filling factors with a compressible 2DES. Remarkably, at filling factor $\nu = 9/2$, where the stripe phase is strongest, the SAW propagation is weakened substantially compared to the overall damping behavior at half filling.

In summary, the pinning mode model presented above predicts a negative velocity change for SAW propagation along the stripe orientation in agreement with our findings. However, it fails to capture some of the experimental observations. In particular, the fact that we observe a negative phase shift over a large frequency range (70 MHz–1 GHz), i.e. also far away from the pinning mode resonance, renders the above interpretation unlikely. It rather points towards the second interpretation outlined below.

#### Negative Electron Compressibility

Microscopically, density-modulated phases are characterized by a clustering of electrons in local domains. This spatial ordering results from a competition between the repulsive and attractive interactions acting among the electrons. It seems natural to associate the presence of an attractive interaction between the electrons with a (dynamic) negative compressibility of the electron system. This term describes the unusual situation that an electron system can lower its chemical potential by increasing the number of charge carriers. Such negative electron compressibilities have been observed in different strongly interacting two-dimensional electron systems [15–18]. A clear signature of negative compressibility is an over-screening of external electric fields.

When studying the propagation of surface acoustic waves, we inherently probe how well the 2DES is able to screen the alternating electric fields created by the SAW. In this respect, the velocity change of the SAW reflects the compressibility of the 2DES. If we now consider a potential negative compressibility of the 2DES in the bubble and stripe phases, the resulting over-screening of the SAW electric field would slow down the sound propagation even further than in the case of a perfect screening. This would cause a negative phase and velocity shift as observed in our experiments. In this picture, the negative phase shift is intimately connected to the strong electron-electron interactions in the bubble and stripe phases. It is important to mention that only the compressibility of the 2DES is probed in the experiment. The overall compressibility of the system remains positive due to the large geometric capacitance between the 2DES and the nearby doping layers.

### *6.3.2 Reentrant States in the Second Landau Level*

In Sect. 2.6 we have pointed out the existence of reentrant states in the second Landau level akin to the bubble phases in higher Landau levels. Hence, it seems obvious to extend the present study also to the reentrant states in the second Landau level. In Fig. 6.3, however, no signatures of these states are observed in $R_{xx}$ and $R_H$. This is first of all a temperature issue. The sample holder used for the SAW experiments is not optimized for obtaining low electron temperatures. Still, to be able to perform SAW measurements also on the fragile reentrant states in the second Landau level, we have fitted suitable coaxial cables with a reduced thermal conductivity to the low-temperature sample holder used in the previous chapters. After improving the thermal anchoring, it was possible to achieve lower sample temperatures. The result is shown in Fig. 6.6. The measurements were done on a different sample with slightly higher quality. In panel a, the longitudinal and Hall resistance is plotted. The transport behavior along the crystal directions [110] and [$1\bar{1}0$] was measured at separate cool-downs. In both directions, the reentrant states in the second Landau level are clearly visible. Remarkably, the easy axis of the stripe phases is rotated compared to the measurements in Fig. 6.3. This rotation of the stripe pattern is a consequence of the slightly lower electron density in the present sample ($2.8 \times 10^{11}$ cm$^{-2}$). As reported by Zhu et al., the stripe phase may interchange its easy and hard axis at a density of about $2.9 \times 10^{11}$ cm$^{-2}$ [19]. The reason for this behavior is unknown so far. Panel b presents the phase shift of the surface acoustic waves. As in Fig. 6.3, a negative phase shift is observed for the bubble and stripe phases. Again, the appearance of

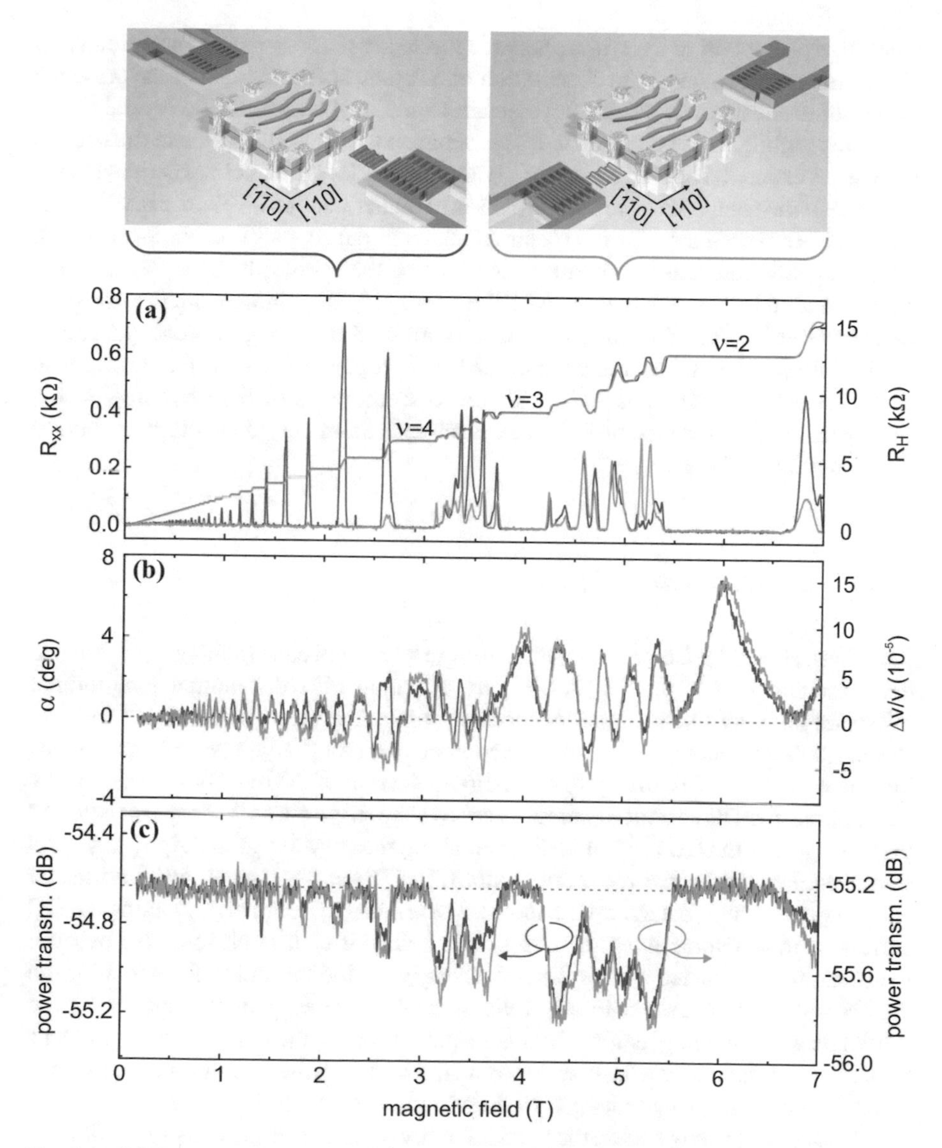


**Fig. 6.6** SAW transmission under improved temperature conditions and sample quality. **a** Longitudinal and Hall resistance for different magnetic fields up to $B = 7\,\mathrm{T}$ measured along the crystal directions [1$\bar{1}$0] (*black*) and [110] (*blue*). The measurements were performed after separate cool-downs. For the stripe phases, the easy axis is rotated compared to Fig. 6.3 due to the lower electron density. **b** SAW phase shift in the same magnetic field range. Strong signatures of the reentrant states in the second Landau level are observed. **c** Power transmitted by the surface acoustic waves. The transmission intensity is enhanced by an external amplifier (factor $10^4$)

a negative phase shift in the stripe phases is limited to the easy axis. Most interesting about this measurement is the observation of a crystal softening also in the presence of the reentrant states in the second Landau level, independent of the crystal direction. This highlights the close connection between the reentrant states in the second Landau level and the bubble phases in higher Landau levels. Panel c shows the SAW power transmission. As for the bubble phases in higher Landau levels, the SAW transmission is enhanced also for the reentrant states in the second Landau level. As an interesting side note, it is important to mention that the bubble phases in the third and higher Landau levels have merged with the nearby IQHS judging from the transport measurement in Fig. 6.6a. Their existence is masked by a single, broad plateau in this low-temperature measurement. For the SAW propagation, however, the presence of the bubble and stripe phases is still well discernible as distinct features, which emphasizes the importance of SAW measurements as an complementary method to study correlated phases.

### 6.3.3 *Impact of current flow*

It has been shown by Göres et al. that a strong DC current can significantly influence the stripe and bubble phases [20]. Figure 6.7 shows the differential longitudinal resistance measured with a small AC current of 10 nA while a DC current of variable strength was sent along the crystal directions [110] and [$1\bar{1}0$]. The DC current was increased in steps of 10 nA up to a maximum current of 300 nA. In the case of the stripe phase, the DC current seemingly destroys the stripe phase if sent perpendicular to the stripes. If the DC current is directed along the easy axis, the stripe phase gets stabilized. For the bubble phase, in contrast, the DC current apparently induces an anisotropic transport behavior with the hard axis being oriented always along the DC current. A microscopic interpretation of this behavior is difficult to derive because the measured resistance values depend strongly on the current path, which in the bubble and stripe phases naturally follows a complicated pattern imposed by the spatial density inhomogeneities. In this respect, probing the conductivity by means of surface acoustic waves is superior to a standard resistive measurement. Surface acoustic waves propagate along a well-defined direction and are only sensitive to length scales much larger than the spatial inhomogeneities of the density-modulated phases.

The corresponding measurements of the SAW propagation are depicted in Fig. 6.8. Shown is the SAW phase shift along both perpendicular crystal axes. The SAW measurements were recorded simultaneously to the transport experiments in Fig. 6.7. Comparing the two figures reveals that the SAW propagation remains conspicuously unaffected by the current flow. This result clarifies that the current flow only locally drives the 2DES out of equilibrium. It is predominantly this local reordering of the stripe pattern which is reflected in the resistive transport measurements. If the current

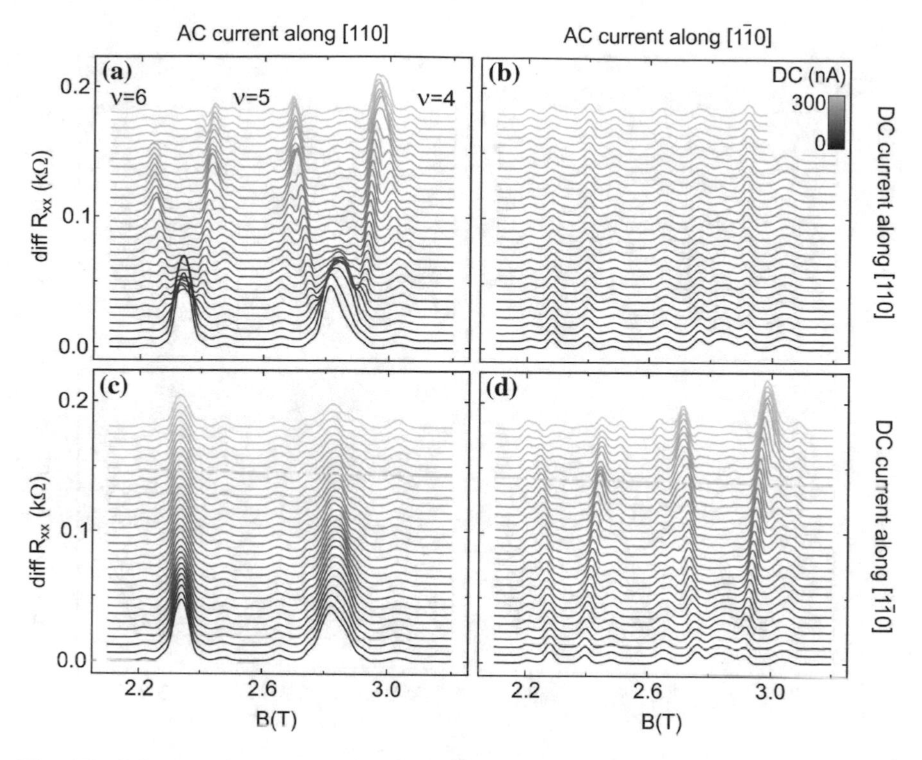


**Fig. 6.7** Influence of a DC current on the differential longitudinal resistance in the density-modulated phases. Plotted is the filling factor range between $\nu = 4$ and 6. A small AC current of 10 nA was used to measure diff $R_{xx}$ along [110] and [1$\bar{1}$0] while the DC current was increased from 0 to 300 nA in steps of 10 nA

is driven orthogonal to the stripe orientation, the stripe order breaks up locally, and a new channel for the current flow is created. The current distribution remains rather well confined along this channel. Away from this region, the stripe pattern remains unaffected. For the isotropic bubble phases, the situation is less clear. Possibly, the strong DC current induces a local stripe order oriented along the Hall field. Such a stripe formation would direct the current flow towards the sample edges and therefore would account for the increased differential resistance observed in Fig. 6.7a, d. Alternatively, the DC current might cause a local depinning or break down of the bubble crystal, which would result in $R_{xx} > 0$. In either case, the reordering of the density distribution must be limited to a small fraction of the sample. Otherwise, signatures of such a substantial structural change would be observable in the SAW measurements.

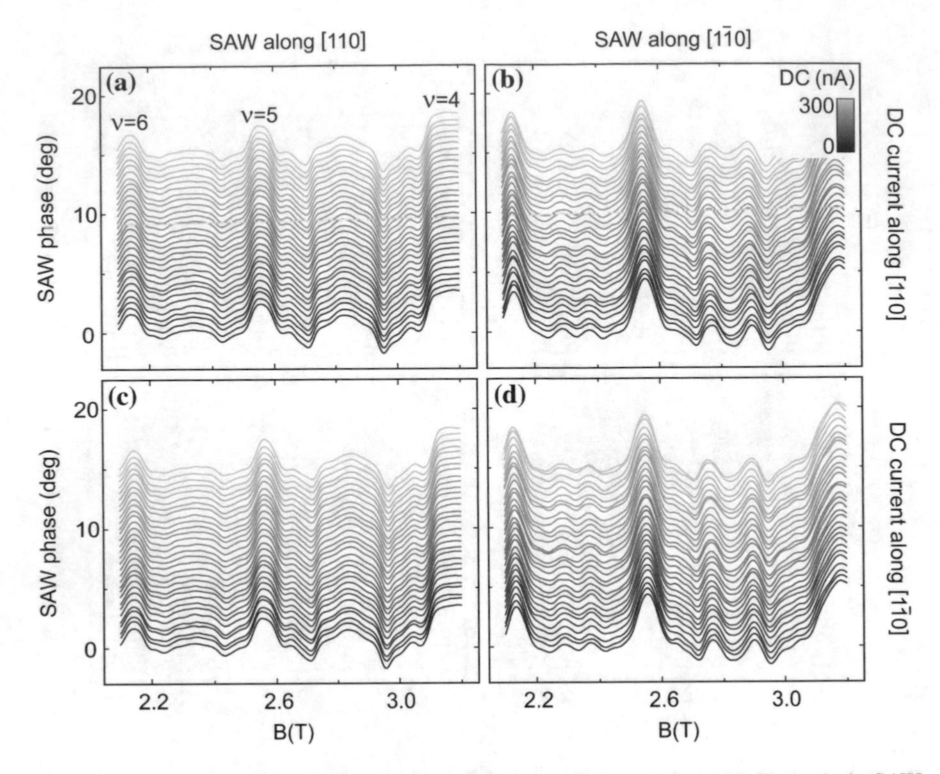


**Fig. 6.8** Influence of a DC current on the propagation of surface acoustic waves. Shown is the SAW phase shift along the crystallographic directions [110] and [1$\bar{1}$0]. The DC current was varied from 0 to 300 nA in steps of 10 nA. The measurements were performed simultaneously to the experiments in Fig. 6.7

# References

1. K.B. Cooper, M.P. Lilly, J.P. Eisenstein, L.N. Pfeiffer, K.W. West, Insulating phases of two-dimensional electrons in high Landau levels: Observation of sharp thresholds to conduction. Phys. Rev. B **60**, 285 (1999)
2. M.P. Lilly, K.B. Cooper, J.P. Eisenstein, L.N. Pfeiffer, K.W. West, Evidence for an anisotropic state of two-dimensional electrons in high Landau levels. Phys. Rev. Lett. **82**, 394 (1999)
3. A. Wixforth, J. Scriba, M. Wassermeier, J.P. Kotthaus, G. Weimann, W. Schlapp, Surface acoustic waves on GaAs/$Al_xGa_{1-x}$As heterostructures. Phys. Rev. B **40**, 7874 (1989)
4. G.W. Farnell, Types and properties of surface waves, in *Topics in Applied Physics: Acoustic Surface Waves* (Springer, Berlin, 1978), pp. 13–60
5. A.R. Hutson, D.L. White, Elastic wave propagation in piezoelectric semiconductors. J. Appl. Phys. **33**, 40 (1962)
6. R. Adler, Simple theory of acoustic amplification. IEEE Trans. Sonics Ultrason. **18**, 115 (1971)
7. A. Wixforth, J.P. Kotthaus, G. Weimann, Quantum oscillations in the surface-acoustic-wave attenuation caused by a two-dimensional electron system. Phys. Rev. Lett. **56**, 2104 (1986)
8. R.L. Willett, M.A. Paalanen, R.R. Ruel, K.W. West, L.N. Pfeiffer, D.J. Bishop, Anomalous sound propagation at $\nu$=1/2 in a 2D electron gas: Observation of a spontaneously broken translational symmetry? Phys. Rev. Lett. **65**, 112 (1990)

9. R.L. Willett, R.R. Ruel, K.W. West, L.N. Pfeiffer, Experimental demonstration of a Fermi surface at one-half filling of the lowest Landau level. Phys. Rev. Lett. **71**, 3846 (1993)
10. R.L. Willett, Surface acoustic wave studies of electron correlations in the 2DES. Surf. Sci. **305**, 76 (1994)
11. S.H. Simon, Comment on "Evidence for an anisotropic state of two-dimensional electrons in high Landau levels". Phys. Rev. Lett. **83**, 4223 (1999)
12. R.L. Willett, K.W. West, L.N. Pfeiffer, Current-path properties of the transport anisotropy at filling factor 9/2. Phys. Rev. Lett. **87**, 196805 (2001)
13. R.M. Lewis, Y. Chen, L.W. Engel, D.C. Tsui, P.D. Ye, L.N. Pfeiffer, K.W. West, Evidence of a first-order phase transition between Wigner-crystal and bubble phases of 2D electrons in higher Landau levels. Phys. Rev. Lett. **93**, 176808 (2004)
14. G. Sambandamurthy, R.M. Lewis, H. Zhu, Y.P. Chen, L.W. Engel, D.C. Tsui, L.N. Pfeiffer, K.W. West, Observation of pinning mode of stripe phases of 2D systems in high Landau levels. Phys. Rev. Lett. **100**, 256801 (2008)
15. S.V. Kravchenko, V.M. Pudaloy, S.G. Semenchinsky, Negative density of states of 2D electrons in a strong magnetic field. Phys. Lett. A **141**, 71 (1989)
16. J.P. Eisenstein, L.N. Pfeiffer, K.W. West, Compressibility of the two-dimensional electron gas: Measurements of the zero-field exchange energy and fractional quantum Hall gap. Phys. Rev. B **50**, 1760 (1994)
17. J.P. Eisenstein, L.N. Pfeiffer, K.W. West, Negative compressibility of interacting two-dimensional electron and quasiparticle gases. Phys. Rev. Lett. **68**, 674 (1992)
18. L. Li, C. Richter, S. Paetel, T. Kopp, J. Mannhart, R.C. Ashoori, Very large capacitance enhancement in a two-dimensional electron system. Science **332**, 825 (2011)
19. J. Zhu, W. Pan, H.L. Stormer, L.N. Pfeiffer, K.W. West, Density-induced interchange of anisotropy axes at half-filled high Landau levels. Phys. Rev. Lett. **88**, 116803 (2002)
20. J. Göres, G. Gamez, J.H. Smet, L. Pfeiffer, K. West, A. Yacoby, V. Umansky, K. von Klitzing, Current-induced anisotropy and reordering of the electron liquid-crystal phases in a two-dimensional electron system. Phys. Rev. Lett. **99**, 246402 (2007)

# Chapter 7
# Summary

More than 30 years after its discovery, the quantum Hall effect keeps surprising us by the stunning richness and complexity of the underlying physical phenomena. Some of the finest pieces of many-body physics are featured in the quantum Hall effect. Its intriguing physics often reappears in other branches of solid state science. The observation of fractionally charged quasiparticles, for instance, is a beautiful manifestation of fractionalization—a many-body phenomenon characterized by the appearance of quasiparticles which cannot be understood as a linear combination of its constituent particles. Similar physics is at play when a single electron decomposes in its fictitious constituents, the spinon, holon and orbiton [1–3], or alternatively in the case of magnetic monopoles in spin ice materials [4–6]. Another beautiful example of quantum Hall physics is the spontaneous symmetry breaking occurring in the density-modulated phases at partially filled Landau levels [7]. It resembles the stripe-shaped charge order known from high-$T_c$ superconductors. Akin to superconductors is also the p-wave pairing of composite fermions, which is believed to take place in the creation of the $\nu = 5/2$ state. Going by this scenario, the $\nu = 5/2$ state would give rise to a fundamental new type of quasiparticles—so-called non-abelian anyons [8, 9]. Their existence would open up a plethora of new experiments. Among these are proposals to utilize the unique braiding statistics of the non-abelian anyons for fault-tolerant topological quantum computation [9, 10]. At this point, the quantum Hall effect connects to the field of quantum information processing. On other frontiers, the quantum Hall effect is a platform to rich spin physics, such as ferromagnetic phase transitions and skyrmionic spin textures [11].

In the light of the magnificent physics transpiring in the quantum Hall regime, we have put the focus on electronic and nuclear spin phenomena as well as the spatial ordering of charges in bubble and stripe phases.

### Electron-Nuclear Spin Interaction in the Quantum Hall Regime

The first experimental chapter deals with the coupling of electronic and nuclear spins in the quantum Hall regime. In GaAs/AlGaAs heterostructures, this coupling is mediated by the contact hyperfine interaction. Its strength can be assessed under the present experimental conditions by measuring the nuclear spin relaxation rate. At



B. Frieß, *Spin and Charge Ordering in the Quantum Hall Regime*,
Springer Theses, DOI 10.1007/978-3-319-33536-0_7

ultra-low temperatures, nuclear spin flips are conveyed preferentially by the coupling to the electron spin system. An important prerequisite to open up this relaxation channel is the presence of low-energetic electron spin excitation in order to bridge the strong imbalance between the electronic and nuclear Zeeman energy. We can access the nuclear relaxation rate by utilizing an electron spin transition at filling factor $\nu = 2/3$ to manipulate the nuclear spin polarization as well as to detect its changes. With this technique at hand, we have investigated systematically the nuclear spin relaxation rate over a large filling factor range. A fast nuclear spin relaxation was found in the vicinity of $\nu = 1$, which is attributed to the presence of skyrmionic spin excitations. At filling factor $\nu = 5/2$, such signatures were conspicuously absent contrary to recent theoretical predictions [12]. Instead, an enhanced nuclear spin relaxation was present at other fractional quantum Hall states. In these cases, a fast spin relaxation was understood as a consequence of composite fermion Landau level crossings. Apart from that, an unexpected and hitherto unobserved enhancement of the relaxation rate was evident at filling factor $\nu = 2$. Moreover, it was found that the equilibrium value of the nuclear spin polarization, i.e. once the nuclei have relaxed completely to thermal equilibrium, depends sensitively on the exact filling factor.

In the second part of this chapter, we studied this filling factor dependence of the nuclear spin polarization in more detail and found a remarkably strong increase of the nuclear spin polarization at filling factors exhibiting a compressible electron system. We were able to largely exclude external influences as the origin of this finding by isolating the interior of the sample from the electrical contacts during relaxation. The observation of a strongly filling-factor-dependent nuclear spin polarization challenges our established understanding of the nuclear spin system in the quantum Hall regime. It would, however, be consistent with a magnetic ordering of the nuclear spins mediated by the Rudermann-Kittel-Kasuya-Yosida (RKKY) interaction. Theoretical calculations predict that the RKKY interaction in a two-dimensional electron system (2DES) can lift the ferromagnetic Curie temperature to the millikelvin range [13, 14]. In the light of future experiments, our findings might constitute the first evidence of a nuclear magnetic ordering under such conditions.

### The Spin Polarization of the 5/2 State

The spin polarization of the $\nu = 5/2$ state has received considerable attention over the last years. It is a crucial quantity to unravel the true nature of the enigmatic $\nu = 5/2$ state. We have approached this question by measuring the characteristic shift of the Zeeman energy experienced by nuclear spins in the presence of a spin-polarized electron system (Knight shift). This shift was investigated by means of nuclear magnetic resonance (NMR) spectroscopy using a resistive detection method. Special attention was attributed to optimize the quality of the $\nu = 5/2$ state as much as possible by increasing the cooling efficiency and reducing the external heat input. In summary, the measured spin polarization at $\nu = 5/2$ is consistent with a completely spin-polarized electron system at all accessible temperatures. This result excludes unpolarized candidate wavefunctions to explain the existence of the $\nu = 5/2$ state, such as the so-called 331 state. Combined with previous measurements of the $e/4$

quasiparticle charge, our findings support the Pfaffian and anti-Pfaffian description of the $\nu = 5/2$ state, both of which predict quasiparticles with non-abelian anyonic exchange statistics.

### Probing the Microscopic Structure of the Stripe Phase at $\nu = 5/2$

The physics of the quantum Hall regime at partial fillings is dominated by the competition between fractional quantum Hall states on the one hand and density-modulated phases on the other hand. This observation becomes most evident in the second Landau level, where density-modulated phases and fractional quantum Hall states are tightly packed and minute changes of the magnetic field or the charge density induce transitions between the respective phases. The close vicinity of density-modulated phases and quantum Hall states is further reflected by the fact that the $\nu = 5/2$ state gives way to a stripe phase when tilting the sample with respect to the external magnetic field.

Our understanding of density-modulated phases originates mostly from their characteristic response to external currents—a reentrance of the integer quantum Hall effect in the case of the bubble phase and strong transport anisotropy in the presence of the stripe phase. Inferred from this transport behavior and substantiated by theoretical calculations, the density-modulated phases are formed by a spatially varying charge density with either a two-dimensional crystalline order (bubble phase) or a one-dimensional stripe order (stripe phase). In this thesis, we have probed for the first time the spatial density distribution of such a stripe phase, more precisely, the stripe phase which emerges at $\nu = 5/2$ when applying an additional magnetic field in the plane of the 2DES. This was achieved by using nuclear spins as local detectors for the electron spin density in the vicinity of each nucleus. Using this technique, we found a remarkably strong modulation of the electron density of about 20 %. Further, it was possible to model the observed nuclear spin response and extract from this the stripe period of the domain pattern.

### Surface Acoustic Wave Study of Density-Modulated Phases

The existence of density-modulated phases in the quantum Hall regime was first experimentally deduced from resistive transport measurements. Since then, experiments of this kind continued to be the method of choice to investigate the stripe and bubble phases in more detail. In the last chapter of this thesis, we have taken a novel approach to probe the conductivity of the density-modulated phases. It utilizes the propagation of acoustic waves along the surface of a sample to study the screening properties of the 2DES below. On piezoelectric substrates, the periodic crystal movement causes an alternating electric field, which in turn influences the sound propagation in the material. At the same time, the electric field interacts with the electrons in the 2DES. This allows us to probe the screening behavior of the 2DES by measuring the propagation velocity of surface acoustic waves (SAWs). We have investigated the SAW transmission in the stripe and bubble phases along both principal axes of the sample. The characteristic transport anisotropy of the stripe phases was found to manifest itself also in the SAW propagation. Surprisingly, if the SAW is traveling along the easy axis of the stripe phase, a reduction of the SAW sound velocity was observed, corresponding to a softening of the crystal. Also for the bubble phases a negative velocity shift is evident but, in contrast to the stripe phase, along both crystal directions. Such a softening of the crystal has not been reported before. It may in the simplest case be understood by the collective excitation of the bubble and stripe crystal in its pinning potential. Driving the system at SAW frequencies higher than the pinning mode resonance might lead to a negative dielectric constant whose magnitude is strong enough to cause a reduction of the SAW velocity. However, additional experiments point towards an alternative interpretation which attributes the observation of a crystal softening in the density-modulated phases to a negative compressibility of the electron system. This unusual property might arise from the strong electron-electron interactions in the bubble and stripe phases.

The clear response of the SAW propagation in the density-modulated phases further allowed clarifying the impact of a strong unidirectional current flow on the stripe phase order. While in resistively detected transport measurements the stripe order appears to be strongly influenced when driving a large current perpendicular to the equilibrium stripe orientation [15], such an effect is conspicuously absent in the SAW measurement. This observation identifies the influence of an external current on the stripe orientation to be purely local. It further highlights the importance of SAW experiments as a complementary method to probe the conductivity.

## References

1. J. Schlappa, K. Wohlfeld, K.J. Zhou, M. Mourigal, M.W. Haverkort, V.N. Strocov, L. Hozoi, C. Monney, S. Nishimoto, S. Singh, A. Revcolevschi, J.-S. Caux, L. Patthey, H.M. Rønnow, J. van den Brink, T. Schmitt, Spin-orbital separation in the quasi-one-dimensional Mott insulator $Sr_2CuO_3$. Nature **485**, 82 (2012)
2. C. Kim, A.Y. Matsuura, Z.-X. Shen, N. Motoyama, H. Eisaki, S. Uchida, T. Tohyama, S. Maekawa, Observation of spin-charge separation in one-dimensional $SrCuO_2$. Phys. Rev. Lett. **77**, 4054 (1996)
3. Y. Jompol, C.J.B. Ford, J.P. Griffiths, I. Farrer, G.A.C. Jones, D. Anderson, D.A. Ritchie, T.W. Silk, A.J. Schofield, Probing spin-charge separation in a Tomonaga–Luttinger liquid. Science **325**, 597 (2009)
4. S.T. Bramwell, S.R. Giblin, S. Calder, R. Aldus, D. Prabhakaran, T. Fennell, Measurement of the charge and current of magnetic monopoles in spin ice. Nature **461**, 956 (2009)
5. T. Fennell, P.P. Deen, A.R. Wildes, K. Schmalzl, D. Prabhakaran, A.T. Boothroyd, R.J. Aldus, D.F. McMorrow, S.T. Bramwell, Magnetic coulomb phase in the spin ice $Ho_2Ti_2O_7$. Science **326**, 415 (2009)
6. D.J.P. Morris, D.A. Tennant, S.A. Grigera, B. Klemke, C. Castelnovo, R. Moessner, C. Czternasty, M. Meissner, K.C. Rule, J.-U. Hoffmann, K. Kiefer, S. Gerischer, D. Slobinsky, R.S. Perry, Dirac strings and magnetic monopoles in the spin ice $Dy_2Ti_2O_7$. Science **326**, 411 (2009)
7. M.M. Fogler, Stripe and bubble phases in quantum Hall systems, in *High Magnetic Fields: Applications in Condensed Matter Physics and Spectroscopy* (Springer, Berlin, 2002), pp. 98–138
8. C. Nayak, S.H. Simon, A. Stern, M. Freedman, S. Das Sarma, Non-abelian anyons and topological quantum computation. Rev. Mod. Phys. **80**, 1083 (2008)
9. A. Stern, Non-abelian states of matter. Nature **464**, 187 (2010)
10. S. Das Sarma, M. Freedman, C. Nayak, Topological quantum computation. Phys. Today **59**, 32 (2006)

11. Y.Q. Li, J.H. Smet, Nuclear-electron spin interactions in the quantum Hall regime, in *Spin Physics in Semiconductors* (Springer, Berlin, 2008), pp. 347–388
12. A. Wójs, G. Möller, S. Simon, N.R. Cooper, Skyrmions in the Moore-Read state at $\nu = 5/2$. Phys. Rev. Lett. **104**, 086801 (2010)
13. P. Simon, D. Loss, Nuclear spin ferromagnetic phase transition in an interacting two dimensional electron gas. Phys. Rev. Lett. **98**, 156401 (2007)
14. P. Simon, B. Braunecker, D. Loss, Magnetic ordering of nuclear spins in an interacting two-dimensional electron gas. Phys. Rev. B **77**, 045108 (2008)
15. J. Göres, G. Gamez, J.H. Smet, L. Pfeiffer, K. West, A. Yacoby, V. Umansky, K. von Klitzing, Current-induced anisotropy and reordering of the electron liquid-crystal phases in a two-dimensional electron system. Phys. Rev. Lett. **99**, 246402 (2007)

# Appendix A
# Sample Parameters

## A.1 Wafer Structure

Most of the samples studied throughout this thesis were developed by Dr. Vladimir Umansky at the Weizmann Institute in Israel. The sample for the measurements presented in Figs. 2.6 and 2.14 was provided by Christian Reichl from the group of Prof. Werner Wegscheider at the ETH Zürich.

All measurements in Chaps. 3–5 were performed on samples of the same wafer. It consists of a GaAs/AlGaAs heterostructure with single-sided doping above the quantum well and an in-situ grown backgate underneath. Its structural details are listed below.

| Quantity | Value |
|---|---|
| Quantum well thickness | 30 nm |
| Depth of the 2DES | 140 nm |
| Spacer thickness | 66 nm |
| Backgate depth | 1064 nm |
| Density at $V_{BG} = 0\,\text{V}$ | $1.8 \times 10^{11}\,\text{cm}^{-2}$ |
| Maximum density | $2.9 \times 10^{11}\,\text{cm}^{-2}$ |
| Maximum mobility | $1.8 \times 10^{7}\,\text{cm}^2/\text{Vs}$ |

Additional details on the conduction band profile and the doping scheme are provided in Sect. 2.1. The measurements in Chap. 6 as well as in Sects. 2.5 and 2.6 were performed on samples without backgate. Instead, these structures had a double-sided doping placed symmetrically on either side of the quantum well. In many cases such structures provide a better quality than equivalent structures with backgate. Yet, for most experiments in this thesis it is required to tune the electron density in a fast and reversible manner.



B. Frieß, *Spin and Charge Ordering in the Quantum Hall Regime*,
Springer Theses, DOI 10.1007/978-3-319-33536-0

## A.2 Device Fabrication

For performing transport experiments, the 2DES was patterned either in the shape of a Hall bar or in a square-shaped van der Pauw geometry. The main steps of the fabrication process are outlined below:

- In the first step, the mesa structure is defined by standard photolithography. For the wet etching, a solution of sulfuric acid and hydrogen peroxide in water is used ($H_2SO_4$:$H_2O_2$:$H_2O$ = 1:8:400). The etch depth is typically around 120 nm. It is chosen such that the upper doping layer is removed in the region outside of the mesa in order to deplete the quantum well underneath.
- For samples with backgate, a separate etch step is required. Here, a typical etch depth is 900 nm. The etching should stop slightly above the backgate layer.
- Before fabricating the contacts to the 2DES, the sample is cleaned in an $O_2$ plasma in order to remove any residues from the etch mask.
- In the next step, the contact material is deposited. It consists of 7 nm Ni, 90 nm Ge, 180 nm Au and 37 nm Ni.
- After lift-off, the contacts are annealed in forming gas at a temperature of roughly 440 °C.
- In the final processing step, the material for the bonding pads is deposited. Here, about 20 nm Cr and 100 nm Au are used.
- For integrating the sample into the measurement setup, the fabricated device is glued onto a 24-pin chip carrier. The electrical contact between chip carrier and sample is established by wire bonding.

# Appendix B
# The Low-Temperature Sample Holder

The physical phenomena investigated throughout this thesis mostly require ultra-low temperatures to be observable. In particular, the $\nu = {}^5\!/_2$ state and the reentrant states in the second Landau level are very fragile. In the course of this Ph.D. work, elaborate measures were taken to optimize the temperature of the sample. The cooling of the sample to temperatures below 20 mK was done using a top-loading $^3$He/$^4$He dilution refrigerator. In this cryostat, the sample is immersed in a mixture of liquid $^3$He/$^4$He. Placing the sample directly in the tail of the mixing chamber ensures a good thermalization of the sample. To minimize the base temperature of the cryostat, it is required to decouple the mixing chamber of the dilution refrigerator as much as possible from the outside world. This reduces the heat input from the warm environment. However, for performing transport and microwave experiments at low temperatures, electrical leads and coaxial cables must be fitted to the sample holder connecting the chip carrier to the measurement equipment outside of the cryostat. In view of these conflicting priorities, it is of highest importance to carefully design the measurement leads and minimize the heat input via these channels. The measures taken in this regard either aim at reducing the heat transfer or improving the cooling efficiency of the sample.

## B.1 Reducing the Heat Load

It seems obvious to choose a material with low thermal conductivity for the 24 measurement wires leading to the sample. However, for most materials this implies also a low electrical conductivity (Wiedemann–Franz law). A class of materials which does not follow this rule is superconductors. They exhibit nearly perfect electron transport combined with a low thermal conductivity. Hence, superconductors serve as an ideal material for the measurement leads. Yet, they can only be used in parts of the sample holder where the temperature is below the transition temperature of the superconductor and where the magnetic field is smaller than the critical field. In our sample holder, the measurement wires are made of a niobium-titanium alloy from



B. Frieß, *Spin and Charge Ordering in the Quantum Hall Regime*,
Springer Theses, DOI 10.1007/978-3-319-33536-0

the 1 K pot down to the tail of the mixing chamber. For the high-frequency coaxial cables, niobium is used as the inner and outer conductor. At parts of the sample holder which do not allow superconducting wires, leads made from stainless steel offer a good compromise for low thermal but high electrical conduction.

### Heat Anchoring

Besides the material choice, it is important to ensure a proper heat anchoring of the wires before entering the mixing chamber. In our system this is done at the 1 K pot. It has a large cooling power and space for additional parts. A thorough pre-cooling is particularly important for the coaxial cables as here the inner conductor is cooled very inefficiently due to the insulating cladding. For this purpose, a printed circuit board with a coplanar waveguide structure is used to bare the inner conductor (Fig. B.1). The board is made of a ceramic with comparably high thermal conductivity.

### Microwave Filtering

In addition to the heat anchoring, it is crucial to install high frequency filters in order to minimize external microwave radiation and thermal noise on the measurement lines. Such filters are particularly useful for the 24 measurement wires. Standard transport measurements are performed under quasi-DC conditions at a modulation frequency below 50 Hz. Hence, no high-frequency transmission is required here. Of course, installing low-pass filters for the coaxial cables would be counter-productive.

An often used high-frequency filter for cryogenic purposes is Thermocoax® cable. This coaxial cable has a nickel-chromium inner conductor and a stainless steel outer conductor. Originally designed as flexible heating wires, Thermocoax® cables turned out to be efficient microwave filters [1]. The high-frequency damping results from losses due to the skin effect. Unfortunately, Thermocoax® wires are difficult to handle since they do not easily wet with solder and because they use MgO as an insulator between inner and outer conductor. MgO is hygroscopic and therefore causes electrical leakage if not sealed properly. We use Thermocoax® wires between the room temperature top of the sample holder and the 1 K pot. In addition, a 3 m long wire is coiled in the mixing chamber part of the sample holder.

**Fig. B.1** Printed circuit board used to thermalize the inner conductor of the coaxial cable

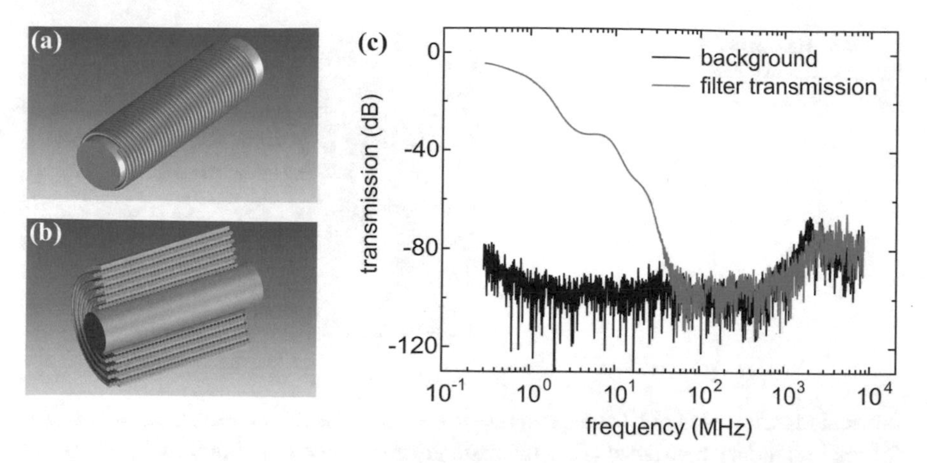


**Fig. B.2** High-frequency filter made of copper wire in a silver epoxy matrix. **a** In the first step, the copper wire is wound around a silver rod. **b** Between consecutive windings silver epoxy is applied such that the copper wire becomes immersed in a silver matrix. **c** Signal transmission through the silver epoxy filter. The background transmission without filter is plotted for comparison

Another high-frequency filter we have worked on is a long copper wire immersed in a matrix of silver epoxy. This type of filter was brought to our attention by Prof. Dominik Zumbühl (Universität Basel) [2]. In a first step, an insulated copper wire with thickness 0.1 mm is wound around a thin silver rod (Fig. B.2a). After completing a full layer, the coil is covered with silver epoxy to completely embed the copper wire in a silver matrix. This entire process is repeated multiple times (Fig. B.2b). In total, the copper wire has a length of roughly 14 m. After completing the filter element, connectors are fitted to either side of the coil such that the transmission line connects to the copper wire. Filters of this kind achieve a remarkable damping behavior. As shown in Fig. B.2c, the signal drops to the noise floor already at frequencies above 100 MHz.

## B.2 Improving the Cooling Efficiency

Reducing the heat input into the mixing chamber is an important prerequisite for cooling samples to ultra-low temperatures. Yet, it does not guarantee low sample temperatures. In fact, very often the temperature of the electrons in the sample is significantly higher than the mixing chamber temperature. This is a consequence of the low cooling efficiency in the millikelvin range. At higher temperatures, the heat transfer is established primarily via phonons. In the millikelvin range, this type of heat transfer is largely suppressed [2, 3]. Instead, at low temperatures the sample is cooled mostly by the conduction electrons in the leads. This raises the question how the leads can be cooled efficiently by the liquid $^3$He/$^4$He mixture. For this purpose, we have

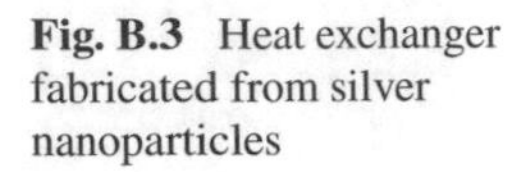

**Fig. B.3** Heat exchanger fabricated from silver nanoparticles

fabricated different heat exchangers, one for each of the 24 measurement wires. The thermal boundary resistance (Kapitza resistance) between the liquid helium and the heat exchanger is inversely proportional to the contact area $\mathcal{S}$ [5]: $R_K \propto T^{-3}\,\mathcal{S}^{-1}$. Hence, a large surface area is required to increase the thermal conductivity. Our heat exchangers are made of silver nanoparticles pressed into a cylindrical shape (Fig. B.3). A sintering of the nanoparticles to increase the stability of the cylinders was not necessary. On either side of the cylinder, a pin is incorporated to connect to the measurement lines. The loose agglomeration of the silver particles brings a large surface area with it, which increases the contact area between the liquid helium and the conductive silver [3–5]. This large interaction surface facilitates an efficient cooling of the electrons.

We have constructed an apparatus which allows us to measure the surface area of these heat exchangers in order to improve the fabrication process and compare different nanoparticle powders. Before presenting the results, we first introduce the underlying measurement principle.

### The BET Method to Determine the Surface Area

The technique used to measure the surface area of the heat exchangers is based on a theory by Brunauer, Emmett and Teller (BET) [6]. It investigates the adsorption of gas molecules on the surface of a nanoporous material. The detailed measurement procedure is described below. All quantities are denoted by a bar to distinguish them from previously used symbols. The measurement setup is sketched in Fig. B.4a.

- At first, the heat exchanger is placed at the bottom of a cylindrical vessel. A vacuum of about $5 \times 10^{-5}$ mbar is pumped while the specimen is heated to a moderate temperature of about 60 °C. During this step, most adsorbates on the surface of the heat exchanger are removed.
- Having reached a decent vacuum, the lower part of the sample vessel is placed in a bath of liquid nitrogen to cool the heat exchanger.
- In the next step, a small amount of nitrogen gas is released from a separate, fixed volume $\overline{V}_1$ into the sample vessel. It is important to measure the pressure $\bar{p}_1$ inside of $\overline{V}_1$ before opening the valve to the sample vessel.

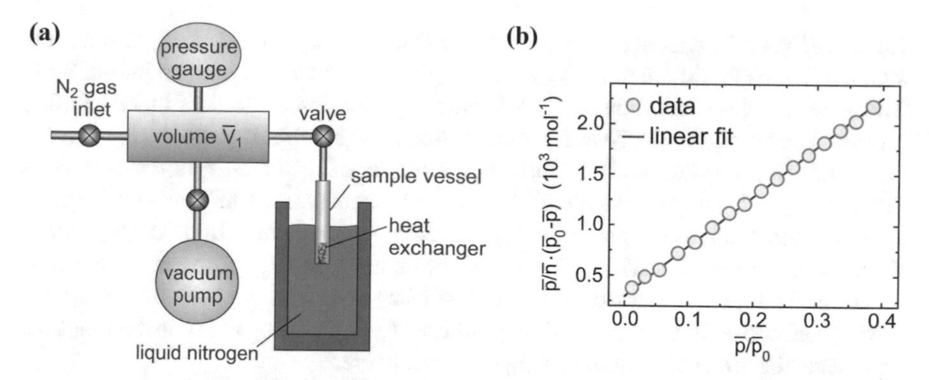


**Fig. B.4** BET method to determine the surface area of nanoporous materials. **a** Schematic of the measurement apparatus. The sample is placed at the bottom of a cylindrical vessel and is cooled from outside with liquid nitrogen. Nitrogen gas is inserted into the sample vessel from a separate dosing volume $\overline{V}_1$. These gas molecules are partly adsorbed on the surface of the sample. **b** The BET theory predicts a linear relation between expression B.2 and the relative pressure $\bar{p}/\bar{p}_0$. From the slope and intersect of the linear fit, the surface area can be calculated

- After opening the valve, the pressure is allowed to settle until an equilibrium is reached. The equilibrium pressure $\bar{p}$ is measured.
- The two previous steps are repeated multiple times while in each cycle nitrogen gas at a slightly higher pressure is inserted into $\overline{V}_1$.

From the gas expansion, the amount of gas adsorbed in each cycle can be calculated. For this purpose, the effective volume $\overline{V}_{eff}$ of the sample cylinder must be known. It is the volume obtained from gas expansion if no heat exchanger is present in the cylinder. It is an effective volume because it considers the entire apparatus to be at equal temperature. The cooler temperatures at the bottom of the sample vessel reduce the pressure inside, which mimics a higher volume. Once $\overline{V}_{eff}$ is known, the adsorbed amount of gas $\Delta\bar{n}$ (in mol) follows directly from the ideal gas equation:

$$\Delta\bar{n} = \frac{\overline{V}_1 \cdot (\bar{p}_1 - \bar{p}) - \overline{V}_{eff} \cdot (\bar{p} - \bar{p}_2)}{\overline{R}\,T}, \tag{B.1}$$

where $\bar{p}_2$ is the pressure in the sample vessel before opening the valve, $\overline{R}$ the universal gas constant and $T$ the temperature of the apparatus (room temperature). For each cycle, the adsorbed gas is calculated according to Eq. B.1. Attention must be paid to keep the liquid nitrogen level outside of the sample cylinder always at the same height. Otherwise, its effective volume would change over time.

The sum of adsorbed gas at equilibrium pressure $\bar{p}$ is labeled $\bar{n}$. According to the BET theory, the expression

$$\frac{\bar{p}}{\bar{n} \cdot (\bar{p}_0 - \bar{p})} \tag{B.2}$$

follows a linear dependence when plotted as a function of $\bar{p}/\bar{p}_0$, where $\bar{p}_0$ denotes the standard atmospheric pressure [6]. An example of such a plot is shown in Fig. B.4b. The measured gas adsorption nicely follows the prediction of the BET theory. Fitting a straight line yields the slope and intercept. According to the BET theory, the amount of $N_2$ gas (in mol) necessary to form a monolayer of adsorbates on the surface is given by $\bar{n}_1 = 1/(\text{slope} + \text{intercept})$. However, a linear relationship between expression B.2 and the relative pressure $\bar{p}/\bar{p}_0$ is only fulfilled for a limited range from $\bar{p}/\bar{p}_0 = 0.05$ to about 0.35 [6]. The exact limits may vary for different nanoporous materials. Having determined $\bar{n}_1$, the total surface area $\mathcal{S}$ of the heat exchanger can be calculated simply by $\mathcal{S} = \bar{n}_1 \, N_A \, \mathcal{S}_N$, where $N_A$ is the Avogadro constant and $\mathcal{S}_N$ represents the area occupied by a single $N_2$ molecule.

## Finding the Optimal Packing Density

We have compared the surface area of heat exchangers made from three different silver nanoparticle powders. The microscopic structure of these powders is shown in Fig. B.5. Powder A is specified for an average particle size of 150 nm by the manufacturer (Inframat® Advanced Materials LLC, USA). Powder B is finer and has an average particle size of 90 nm (M K Impex Corp., Canada). Powder C consists of the smallest particles and is listed with an average particle size of about 60 nm (Inframat® Advanced Materials LLC, USA). Not only the particle size is visibly smaller for this powder but also the variation in size appears to be reduced. The three powders were pressed into cylindrical heat exchangers (Fig. B.3). The cylinders have a diameter of 8 mm and a height of 21 mm. The dimensions are optimized for the spatial constraints in the sample holder. Shortly before pressing the heat exchangers, the powder was treated with forming gas for one hour at a temperature of 60 °C in order to clean the particle surface [4, 5].

For the given heat exchanger dimensions, we have studied systematically how the surface area changes for different packing densities. It was possible to vary the density from 30 to 80 % relative to the density of massive silver. For smaller densities, the powder is only loosely connected and does not stay in shape. For higher densities, the

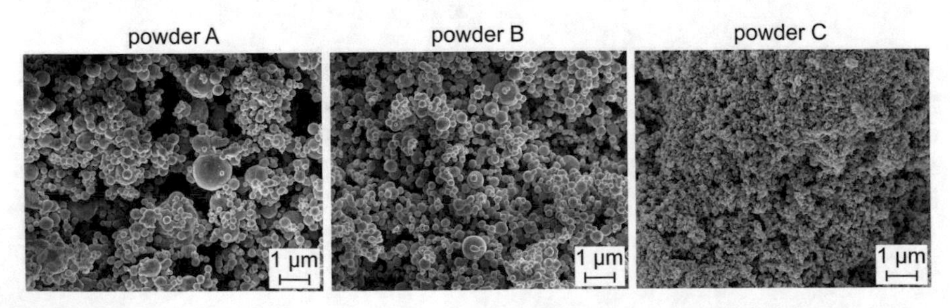


**Fig. B.5** Microscopic image of powder A, B and C with an average particle size of 150, 90 and 60 nm, respectively

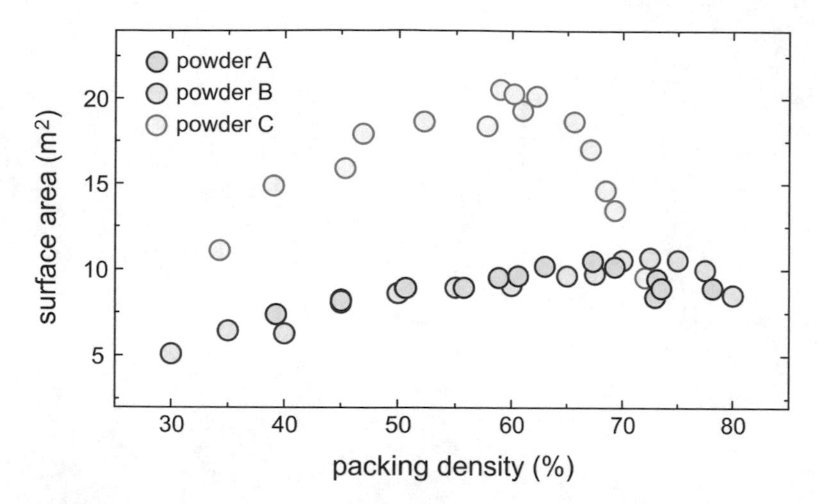


**Fig. B.6** Surface area of heat exchangers made of different powders A, B and C with varying packing densities

molding press gets damaged. The measured surface areas are depicted in Fig. B.6. Powder C, the powder with the smallest particles, yields the largest surface area. Despite the small particle size, a widely ramified network of nanoscale pores must have formed during pressing such that the overall surface area becomes larger. At a density of 60 %, the surface becomes maximal and has a value of roughly 20 $m^2$. When going to even higher densities, the size and number of pores gets reduced, causing a smaller surface area.

## References

[1] A.B. Zorin, The thermocoax cable as the microwave frequency filter for single electron circuits. Rev. Sci. Instrum. **66**, 4296 (1995)

[2] C.P. Scheller, S. Heizmann, K. Bedner, D. Giss, M. Meschke, D.M. Zumbühl, J.D. Zimmerman, A.C. Gossard, Silver-epoxy microwave filters and thermalizers for millikelvin experiments. Appl. Phys. Lett. **104**, 211106 (2014)

[3] F. Pobell, *Matter and Methods at Low Temperatures* (Springer, Berlin, 2007)

[4] H. Franco, J. Bossy, H. Godfrin, Properties of sintered silver powders and their application in heat exchangers at millikelvin temperatures. Cryogenics **24**, 477 (1984)

[5] W. Itoh, A. Sawada, A. Shinozaki, Y. Inada, New silver powders with large surface area as heat exchanger materials. Cryogenics **31**, 453 (1991)

[6] S. Brunauer, P.H. Emmett, E. Teller, Adsorption of gases in multimolecular layers. J. Am. Chem. Soc. **60**, 309 (1938)

# Curriculum Vitae

### Current employment

since 05/2010 Research scientist at the Max Planck Institute for Solid State Research in Stuttgart, group of Prof. Dr. Klaus von Klitzing

09/14 Doctorate at the Technical University of Munich (summa cum laude)

### Higher education

10/04–10/09 Physics course (honors program) at the University of Würzburg

10/2009 Master of Science with honors

11/2007 Bachelor of Science

### Practical experience

07/03–09/12 Student employee at the Institute for Radiotherapy in Schweinfurt

01/10–03/10 Research engineer at Robert Bosch GmbH in Gerlingen

08/08–10/09 Master thesis at the University of Würzburg

01/09–04/09 Research project at the Stanford University (USA)

03/08–06/08 Research project at the University of Oxford (UK)

09/07–10/07 Bachelor thesis at the University of Würzburg

02/07–04/07 Research internship at the MPI for Metals Research in Stuttgart

02/06–04/06 Internship at Siemens Medical Solutions in Forchheim

### Scholarships and awards

06/2015 Otto Hahn Medal

03/07–10/09 Scholarship of the German National Academic Foundation

01/09–04/09 Scholarship of the German Academic Exchange Service

01/2005 Moral courage award



B. Frieß, *Spin and Charge Ordering in the Quantum Hall Regime*,
Springer Theses, DOI 10.1007/978-3-319-33536-0

## Publications

### Articles

- J. Falson, D. Maryenko, B. Friess, D. Zhang, Y. Kozuka, A. Tsukazaki, J.H. Smet, M. Kawasaki, Even-denominator fractional quantum Hall physics in ZnO. Nat. Phys. **11**, 347 (2015)
- T. Guan, B. Friess, Y.Q. Li, S.S. Yan, V. Umansky, K. von Klitzing, J.H. Smet, Disorder-enhanced nuclear spin relaxation at filling factor one. Chin. Phys. B. **24**, 067302 (2015)
- B. Friess, V. Umansky, L. Tiemann, K. von Klitzing, J.H. Smet, Probing the microscopic structure of the stripe phase at filling factor 5/2. Phys. Rev. Lett. **113**, 076803 (2014)
- J. Nuebler, B. Friess, V. Umansky, B. Rosenow, M. Heiblum, K. von Klitzing, J. Smet, Quantized $\nu$=5/2 state in a two-subband quantum Hall system. Phys. Rev. Lett. **108**, 046804 (2012)
- T.D. Ladd, D. Press, K. De Greve, P.L. McMahon, B. Friess, C. Schneider, M. Kamp, S. Höfling, A. Forchel, Y. Yamamoto, Nuclear feedback in a single electron-charged quantum dot under pulsed optical control, in *Proceedings of the SPIE79480U* (2011)
- D. Press, K. De Greve, P. McMahon, T. Ladd, B. Friess, C. Schneider, M. Kamp, S. Höfling, A. Forchel, Y. Yamamoto, Ultrafast optical spin echo in a single quantum dot. Nat. Photonics **4**, 367 (2010)
- T. Ladd, D. Press, K. De Greve, P. McMahon, B. Friess, C. Schneider, M. Kamp, S. Höfling, A. Forchel, Y. Yamamoto, Pulsed nuclear pumping and spin diffusion in a single charged quantum dot. Phys. Rev. Lett. **105**, 107401 (2010)

### Conference Contributions

- Invited presentation: Spin and charge ordering in the quantum Hall regime, *International Workshop on Emergent Phenomena in Quantum Hall Systems (EPQHS)*, Mumbai, India (2016)
- B. Friess, Y. Peng, B. Rosenow, F. von Oppen, V. Umansky, K. von Klitzing, J.H. Smet, Negative compressibility of the bubble and stripe phases in the quantum Hall regime, oral presentation, *International Symposium on Advanced Nanodevices and Nanotechnology (ISANN)*, Hawąii, USA (2015)
- Invited presentation: NMR probing of the spin and charge ordering in the quantum Hall regime, *International Workshop on Quantum Nanostructures and Electron-Nuclear Spin Interactions*, Tohoku Forum for Creativity, Sendai, Japan (2015)
- B. Friess, B. Rosenow, V. Umansky, K. von Klitzing, J.H. Smet, Negative compressibility of the bubble and stripe phases in the quantum Hall regime, oral presentation, *International Conference on Electronic Properties of Two-Dimensional Systems (EP2DS)*, Sendai, Japan (2015)
- B. Friess, V. Umansky, L. Tiemann, K. von Klitzing, J.H. Smet, Probing the stripe phase at filling factor 5/2 by NMR, poster presentation, *International Workshop on Emergent Phenomena in Quantum Hall Systems (EPQHS)*, Rehovot, Israel (2014)

- B. Friess, V. Umansky, L. Tiemann, K. von Klitzing, J.H. Smet, Probing the stripe phase at filling factor 5/2 by NMR, oral presentation, *International Conference on Electronic Properties of Two-Dimensional Systems (EP2DS)*, Wroclaw, Poland (2013)
- B. Friess, V. Umansky, K. von Klitzing, J.H. Smet, Electronic and nuclear spin properties in the quantum Hall regime, oral presentation, *Workshop on Topological Quantum Computation*, Leipzig, Germany (2012)
- B. Friess, V. Umansky, K. von Klitzing, J.H. Smet, Probing the spin polarization of the $\nu$=5/2 fractional quantum Hall state, oral presentation, *Workshop on Topological Materials for Nanoelectronics*, Stuttgart, Germany (2012)
- B. Friess, V. Umansky, K. von Klitzing, J.H. Smet, Filling factor dependence of electron-nuclear spin interaction, poster presentation, *International Winterschool on New Developments in Solid State Physics*, Mauterndorf, Austria (2012)

Printed in the United States
By Bookmasters